The Curious Hats of Magical Maths

Excerpts from Reviews

J.T. Glover...shows clearly that when the formulas are understood, particularly in relation to one another, Vedic mathematics presents unified mathematics.

—*Journal of Oriental Research, Chennai, Vol. 71-73, 2000-2003*

The book has been neatly printed and moderately priced. For its immense intellectual worth, it is highly recommended to the seekers of knowledge of mathematics.

— *Vedic Astrology, Vol. 6, No. 4, July-August, 2002*

It is totally different from the usual textbooks we see at the school level. Written in simple English, such a book should find a place in every school library.

—*The Vedanta Kesari, September, 2004*

Vedic Mathematics throws open a welcome opportunity for mathelets or whizkids to hone their skills and win laurels through display of mental feat in giving one-line quick responses to mathematical calculations and manipulations.

— *The Hindu, September 5, 2000*

The examples and exercises have been arranged with care and the grading of the latter shows every evidence of the same pedagogic thoroughness. ...For this reason, if no other, the book deserves to serve as a model for text books at this level, in use in this country.

— *The Hindu, February 27, 1996*

As against attempts by earlier writers, the presentation, the illustrations and the exercises are all executed very well. The text is interspersed with some geometrical constructions and graphs presumably to answer the requirements of the system.

—*Bhavan's Journal, Vol. 46, No. 14, February, 2000*

J.T. Glover rightly states, mathematics has two directions—an outer and an inner. The outer direction is fulfilled by applying the mathematical formulae. The inner direction is the investigation of the real nature of the working phenomena from the origin of man or universe up to the ultimate reality. There are many more interesting and intelligent short-cuts in multiplication, division and so on. This book also demonstrates how 'Vedic Mathematics belongs not only to an hoary antiquity but also is... as modern as the day after tomorrow.'

—*S.V.U. Oriental Journal, Vol. xliii*

The Curious Hats of Magical Maths

Vedic Mathematics for Schools

Book 1

JAMES GLOVER

**MOTILAL BANARSIDASS PUBLISHERS
PRIVATE LIMITED ● DELHI**

Revised and Enlarged Edition: Delhi, 2015
First Edition : Delhi, 1999 **(reprinted seven times)**
Published under the title "Vedic Mathematics for Schools"

© JAMES GLOVER
All rights reserved

Book 1 : ISBN: 978-81-208-3973-1 (PB)
ISBN: 978-81-208-3991-5 (HB)
Book 2 : ISBN: 978-81-208-3974-8 (PB)
ISBN: 978-81-208-3992-2 (HB)

MOTILAL BANARSIDASS
41 U.A. Bungalow Road, Jawahar Nagar, Delhi 110 007
8 Mahalaxmi Chamber, 22 Bhulabhai Desai Road, Mumbai 400 026
203 Royapettah High Road, Mylapore, Chennai 600 004
236, 9th Main III Block, Jayanagar, Bengaluru 560 011
8 Camac Street, Kolkata 700 017
Ashok Rajpath, Patna 800 004
Chowk, Varanasi 221 001

Printed in India
by RP Jain at NAB Printing Unit,
A-44, Naraina Industrial Area, Phase I, New Delhi–110028
and published by JP Jain for Motilal Banarsidass Publishers (P) Ltd,
41 U.A. Bungalow Road, Jawahar Nagar, Delhi-110007

Dedication

For all who love numbers, have an open mind and in whom the search for knowledge shines in their hearts.

Author's Preface

This book is designed for young and old who might enjoy learning and practising the Vedic methods of Mathematics. Vedic mathematics is unconventional and not very well known and so readers are invited to be open-minded in their approach. The system uses the nature of number and natural mental processes to provide quick and easy methods for all sorts of calculations. Many difficult-looking problems can be solved at lightning speed with the answer coming digit-by-digit and without stress or anxiety. When practised the methods give great delight and a sense of the magical quality of numbers. The amazing simplicity and wonderment of obtaining the answers so easily has led some to ask, is this maths or magic? The answer is that it is magic until you have understood how it works and thereafter it is both maths and magic.

It has been nearly twenty years since I wrote Vedic Mathematics for Schools Books 1, 2 and 3 and have long felt that a completely new format is needed. In those previous books I attempted to cover, from a Vedic standpoint, the topics common to many schools teaching maths to 11 - 13 year-olds. In these second editions my aim is just to introduce the main topics of Vedic maths relevant to children of a similar age and which can be used as either support or extension material for teachers or just for interest by anybody. Books 1 and 2 are designed to be workbooks in which there are spaces to write the answers. Book 1 covers the main basic methods of calculation which use the Vedic rules. They are not blanket rules as there are plenty of methods used for specific cases as well as covering the general case. Vedic mathematics is highly flexible in two senses. Firstly, there are often several ways to get to an answer, all of which are entirely valid and correct and the reader can then have the flexibility of choosing whichever one seems most appropriate. Secondly, each of the rules has many varied and different applications and uses. They are flexible in themselves.

Some of the Vedic rules are short yet fairly cryptic. I had the idea of representing each sutra by a hat with a particular design. Hence the title of the book. The design of each hat in some way reflects the meaning of the rule.

It is my sincere hope that, in working through this book you will come to enjoy and love working with number and with the sutras.

I am greatly indebted to my daughters Sophie and Amy - Sophie for designing and producing the wonderful illustrations for the hats as well as the cover design and Amy for her help in checking the answers. I would like to thank them for their support and assistance.

James Glover September 2013

Contents

Introduction	vi
Chapter 1 *Nikhilam* **Multiplication**	1
Chapter 2 Complements	8
Chapter 3 The Deficiency	14
Chapter 4 Number Patterns	18
Chapter 5 Multiplying above the base	25
Chapter 6 How many numbers are there?	30
Chapter 7 Multiplication by *Vertically and Crosswise*	35
Chapter 8 Subtraction by *On the Flag*	42
Chapter 9 *Nikhilam* **Subtraction**	46
Chapter 10 Nikhilam Division	53
Chapter 11 *Proportionately*	63
Chapter 12 The meaning of number	68
Chapter 13 Algebra and equations	72
Chapter 14 A few short cuts for multiplying	82
Chapter 15 Digital Roots	86
Chapter 16 Further steps with *Nikhilam* **multiplication**	94
Chapter 17 Working with decimal fractions	98
Chapter 18 Puzzles and problems	104
Appendix The Sutras of Vedic Mathematics	112
Answers	115

Introduction

The Curious Hats of Magical Maths is an introductory workbook on Vedic Mathematics. It leads you into some unique, enjoyable and very quick methods of working with numbers. There are full descriptions of these methods together with worked examples for you to follow and plenty of practice exercises. This is a workbook and you can write your answers into the spaces provided. You will find the answers at the back of the book to check your work. The problems and methods are suitable for any age but probably most apt for 11 – 13 year-olds, Indian Class VI – VII, UK Years 7 – 8. The aim is to introduce some of the Vedic mathematical techniques and not to cover the whole of the mathematics syllabus for 11 - 13 year-olds. So the book can be used for support material, extension material or simply by those wishing to find fast techniques for solving problems.

Veda

Vedic Maths is a system from India and, in modern times, was first written about by a spiritual teacher named Bharati Krishna Tirtha. Mathematics was his hobby and he discovered the system from his own studies of ancient teachings. The word *Veda* means knowledge and so Vedic Maths means *Knowledge Maths*. It is based on sixteen simple rules called sutras (pronounced "sootras") and a sutra is a thread of knowledge. Veda is also as the name for the ancient teachings of India, thousands of years old, which deal with all manner of aspects of life, both spiritual and worldly. In fact, this is the most common use of the name. Traditionally, these teachings were handed down by word of mouth and learnt by heart. This is called an oral tradition. It is therefore not possible to know their exact age. It is also possible that not all of these teachings are published anywhere. Be that as it may, Tirtha was a brilliant scholar and an inspiring mathematician. He left behind one volume giving illustrative descriptions of some applications of the sutras first published in 1965. Since then there has been an increasing interest in his system of Vedic mathematics and there are now several websites available to learn more about it.

Sutras

The sixteen sutras of Vedic mathematics are short, easily memorised, statements giving principles, patterns of working or rules of thumb for solving all sorts of mathematical problems by the fastest and easiest routes. The aim of the sutras is to provide easy methods involving mental working. For the most part each sutra covers a wide range of topics and this book deals with introductory applications which are then further developed in Book 2.

This book introduces you to methods of multiplication, division and subtraction using the Vedic sutras as well as other aspects of arithmetic and algebra. Many of the methods will be new to you and others you may know. There are many short cuts in maths which will be natural to you. For example, if you mentally add 465 and 299, most will find the easiest way is to add 300 and take 1 off, leaving 764. This is quite natural and not unknown. You will have used a deficiency, the fact that 299 is 1 less than 300. The Vedic sutras follow these natural processes and point them out. So there are special methods as well as general methods

The last chapter provides puzzles and problems which can be solved using the sutras. There may be nothing out of the ordinary in the way you solve these problems because the sutras are quite natural. You can answer the questions and the sutras indicate the way you think about the problems.

Numbers

Mathematics is based on number and numbers begin with unity at 1. The numbers are 1 to 9 and these, together with zero, make up all the numbers we use in everyday life. If you treat these as ten friends then there is no need to fear them. Just like good friends they are completely reliable and trustworthy. They do not change with time. No matter how big a number is, it is always made up of these nine and the zero. This is described by means of a story in Chapter 6. The symbols we all use for numbers have become universal. All children throughout the world learn them. But they originate, including the zero, from thousands of years ago in India. It is therefore fitting that this wonderful system of Vedic mathematics also comes from India. No one knows why there are nine numbers and a zero and that we use a base ten for our number system. Some say that it is because we have ten fingers on our hands. Others say that it follows an ancient description of the creation involving nine elements. In Chapter 12 is another story briefly describing these elements.

Magic Hats

Each sutra is represented as a 'magic hat'. Just as a nurse wears a hat or a jockey wears a hat which helps them think and act in a way suitable for their work so you can pretend to wear each magic hat to help remind you of the way to think.

The hats give a visual representation of the sutras.

The Curious Hats

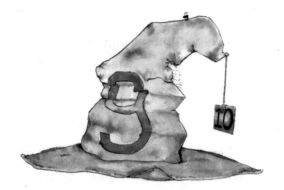

All from 9 and the last from 10

By one more than the one before

Vertically and Crosswise

The Deficiency

By Addition and Subtraction

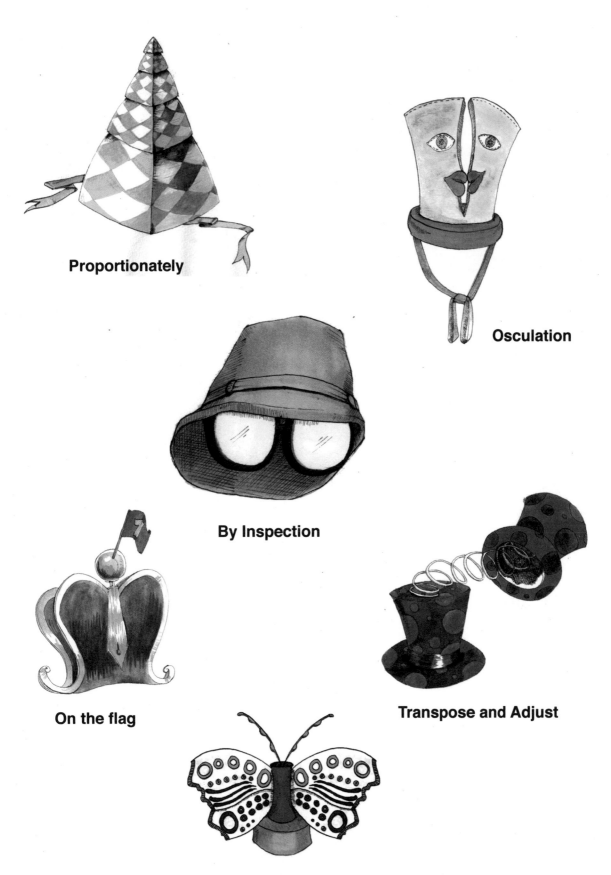

Proportionately

Osculation

By Inspection

On the flag

Transpose and Adjust

When the total is the same, it is nought

Chapter 1

Nikhilam Multiplication

Nikhilam multiplication is a wonderfully quick and easy method for multiplying numbers close to a power of ten. The reason it is called *Nikhilam* will become clear in a later chapter but it is an abbreviation for the rule, or sutra, describing the method.

There are many calculations in arithmetic which are made easier using a complement or deficiency. For example, if you want to add 257 and 99 you may find the easiest way is to add 100 and take off 1 leaving 356. 99 is deficient from 100 by 1 and 1 is called the complement of 99. This method uses complements.

Base 10

Example 7×8

1. 10 is the base because it is the nearest unity to 7 and 8 and this is written above as a reminder. Take 7 and 8 away from 10 to find the complements 2 and 3. Put these down on the right-hand side with a connecting minus sign.

$$\begin{array}{r} 10 \\ 7-3 \\ \times\, 8-2 \\ \hline /\ \end{array}$$

The minus sign shows that the complements are both less than 10. The stroke is used to distinguish the two parts of the answer.

$$\begin{array}{r} 10 \\ 7-3 \\ \times\, 8-2 \\ \hline 5\,/\,6 \end{array}$$

2. On the right-hand vertically multiply the two complement digits, $3 \times 2 = 6$ and write it down.

3. For the left-hand cross-subtract, either $7 - 2 = 5$ or $8 - 3 = 5$. Both give the same answer. Write that down on the left. So the final answer is 56.

1

The Curious Hats of Magical Maths

This method holds good in all cases but is only easy when both or one of the numbers is close to the same base. It is said that a very long time ago, the cross-subtraction part gave rise to the × sign being used for multiplication.

1.1 *Use this method even though you might know the answers already*

1. 9
    ×9

2. 8
    ×9

3. 7
    ×9

4. 6
    ×9

5. 5
    ×9

6. 4
    ×9

7. 1
    ×9

8. 3
    ×9

1.2 *Here are some more examples*

1. 2
    ×9

2. 9
    ×8

3. 8
    ×8

4. 7
    ×8

5. 6
    ×8

6. 9
    ×7

7. 8
    ×7

8. 7
    ×7

9. 9
    ×6

10. 8
    ×6

11. 9
    ×5

12. 9
    ×4

13. 9
    ×3

14. 9
    ×2

15. 9
    ×1

16. 6
    ×9

When using complements you relate each number to 10 and 10 is a unity. By doing multiplication like this you have a way of calculating the times tables further than . Of course you should learn these tables otherwise mathematics becomes difficult later on.

Chapter 1 Nikhilam Multiplication

Using a base of 100

What happens when the numbers are larger? The next example shows how to use 100 as the base. Both numbers are a little less than this. 100 has two zeros and so the complement has two digits. Apart from this the method is exactly the same as before.

Example 94×97

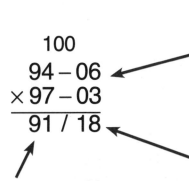

1. Find the complements of 94 and 97, 06 and 03. Two digits are used as indicated by the number of zeros in 100.

2. Again, the answer comes in two parts. On the right-hand side multiply the complements, $6 \times 3 = 18$.

3. On the left, cross-subtract as before. This gives $94 - 3 = 91$ or $97 - 6 = 91$. Remember you can cross-subtract either way, whichever is easier. The answer is 9118.

1.3 Try these

1.	95 × 97	4.	98 × 96	7.	91 × 93	10.	94 × 96
2.	97 × 97	5.	95 × 94	8.	91 × 95	11.	95 × 97
3.	87 × 97	6.	94 × 92	9.	92 × 97	12.	98 × 94

The Curious Hats of Magical Maths

1.4 *Further practice*

1.	87 × 98	5.	89 × 93	9.	77 × 98	13.	96 × 96
2.	92 × 93	6.	68 × 98	10.	91 × 92	14.	95 × 91
3.	79 × 97	7.	67 × 99	11.	67 × 97	15.	85 × 96
4.	88 × 94	8.	86 × 97	12.	72 × 99	16.	94 × 90

1.5 Work these out then shade in the answers in the grid. Find the mystery number.

92 × 97	46 × 99	93 × 95	93 × 99
93 × 94	92 × 93	52 × 99	96 × 92
91 × 94			93 × 91
90 × 98			88 × 96
39 × 99	93 × 93	71 × 99	93 × 96

9889	8832	8765	9207	8556	9773
8097	8924	8765	4976	9977	9624
8967	7029	8463	8835	3861	8945
8758	8879	5774	6992	5148	7889
8845	8820	9687	4671	4554	9580
9034	8649	8554	8448	8742	3289

Chapter 1 Nikhilam Multiplication

Carrying to the left

When using a base of 100 you are allowed only two digits on the right-hand side of the answer. The guide for this is the number of zeros in the base 100, namely, two. Here is a rhyme to help you remember.

The number of digits in the complement's place
is the same as the number of noughts in the base.

Example 76×92

$$\begin{array}{r} 76-24 \\ \times 92-08 \\ \hline 69/_19_32 \end{array}$$

1. Using All from 9 and the last from 10, the complements are 24 and 8.

2. On the right multiply $4 \times 8 = 32$. Put down 2 and carry 3.
Multiply $2 \times 8 = 16$, and add in the carry 3. Put down 9 and carry 1.

3. On the left cross subtract, $76 - 8 = 68$, and add the carry 1 to make 69.
The final answer is 6992.

1.6 *With carrying*

1.	79 × 94	5.	79 × 92	9.	19 × 98	13.	57 × 92
2.	69 × 95	6.	79 × 91	10.	49 × 97	14.	47 × 91
3.	59 × 96	7.	78 × 95	11.	68 × 95	15.	86 × 91
4.	79 × 93	8.	75 × 92	12.	77 × 94	16.	75 × 95

5

The Curious Hats of Magical Maths

1.7 *Further practice*

1. 67 × 93	5. 68 × 94	9. 67 × 95	13. 37 × 95
2. 69 × 92	6. 78 × 94	10. 58 × 95	14. 66 × 93
3. 85 × 92	7. 67 × 94	11. 78 × 92	15. 36 × 97
4. 88 × 90	8. 77 × 95	12. 68 × 94	16. 43 × 97

This method is called **Nikhilam multiplication**. This is because the first word of the rule, in its original language is *Nikhilam*. In Sanskrit, the rule is *Nikhilam navatascaraman dasatah* and means *All from 9 and the last from 10*. It's a very fast method when one or both the numbers are close to a base.

1.8 *See how fast you can do these*

1. 26 × 99	4. 94 × 98	7. 12 × 99	10. 77 × 98
2. 73 × 99	5. 39 × 99	8. 17 × 99	11. 68 × 98
3. 48 × 99	6. 75 × 98	9. 79 × 97	12. 67 × 97

Chapter 1 Nikhilam Multiplication

This is a special method and can be used when the numbers are reasonably close to a base. When this is not the case the method can be tricky. For example, if you were to multiply 26 by 72 using this method the complements would be 74 and 28. The multiplying of complements on the right would then be at least as difficult as what you started with. Fortunately, the general method of multiplying by *Vertically and Crosswise* takes care of this. You will meet this in Chapter 7. You will be able to multiply any two numbers together in one line of easy quick working. Multiplying numbers above the base is dealt with in Chapter 5.

How will you know which method to use? This comes from experience. For now it is important to practice so as to get really used to the method.

1.9 Revision

1. 98 × 97
2. 92 × 98
3. 93 × 94
4. 98 × 98
5. 56 × 97
6. 88 × 88
7. 96 × 95
8. 91 × 92
9. 94 × 94
10. 93 × 98
11. 83 × 92
12. 89 × 89
13. 91 × 99
14. 92 × 97
15. 88 × 96
16. 78 × 98
17. 43 × 98
18. 37 × 98
19. 89 × 96
20. 92 × 92
21. 91 × 95
22. 62 × 99
23. 70 × 92
24. 49 × 94

Chapter 2

Complements

In this chapter you will be learning how to find the complement of a number. A complement is what must be added to a number to make it up to the next 10, 100, 1000, and so on. So for 10 the complement of 7 is 3 because 7 + 3 = 10. For 100, the complement of 97 is 3 because 97 + 3 = 100. A complement is a deficiency from a power of 10.

The easy way to find the complement of a number from 100 is to take the first number away from 9 and the last number away from 10. For example, for 86, 8 from 9 is 1 and 6 from 10 is 4. Putting these two together makes 14 and so 14 is the complement of 86.

This means that 86 + 14 = 100 and 100 − 14 = 86.

1. Take this number from 9 to obtain 1. → 86 ← *2. Take the last number from 10.*
　　　　　　　　　　　　　　　　　　14

14 is the complement of 86. 86 is also the complement of 14.

9 Complements		10 Complements	
0	9	0	10
1	8	1	9
2	7	2	8
3	6	3	7
4	5	4	6
		5	5

Chapter 2 Complements

2.1 Write the complement of each of the following numbers underneath

1. 64   46   52   99   37   12

2. 24   83   67   48   51   62

3. 33   47   72   56   85   21

4. 45   98   23   34   59   76

5. 93   72   86   65   49   32

6. 44   36   22   38   13   68

All from 9 and the last from 10

All from 9 and the last from 10 is the rule which tells you how to find the complement of any number. Here is its cap!

To find the complement of a number like 876 you can take each number away from 9 and the last from 10 to give 123.

$$\begin{array}{c} 8\ 7\ 6 \\ 1\ 2\ 3 \end{array}$$

123 is the complement of 876 because when added together they make 1000.

$$\begin{array}{r} 8\ 7\ 6 \\ +\ 1\ 2\ 3 \\ \hline 1_1 0_1 0_1 0 \end{array}$$

9

The Curious Hats of Magical Maths

2.2 Write the complement of each number underneath

1. 978   765   456   302   766   284

2. 452   308   761   859   774   623

3. 987   869   123   666   705   647

4. 7684   4656   9032   5647   1003   3777

5. 8609   2045   5007   8855   3333   8993

6. 1212   1158   8303   5742   4079   2103

What to do when the number ends in zero

The rule has the meaning, *All from nine and the last **number** from ten*. Since zero is not a number you only take the last number from 10.

Example Find the complement of 6200.

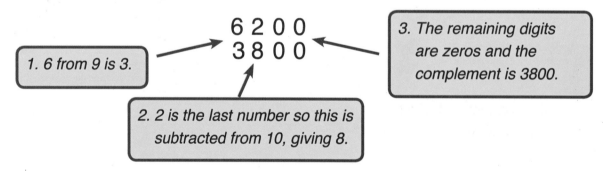

Example Find the complement of 2040.

1. 2 from 9 is 7, 0 from 9 is 9.

2. 4 is the last number, so it's taken from 10. The complement is 7960.

Chapter 2 Complements

2.3 Write the complement of each number underneath

1. 40   70   160   80   60   90

2. 420   750   390   630   210   780

3. 600   300   400   900   200   700

4. 4650   7680   2150   5470   9080   3050

2.4 Further practice

1. 12   60   17   16   69   42

2. 504   604   234   501   21   562

3. 9999   123   7654   1098   83   280

4. 5687   3986   58700   11102   60604   30500

5. 3257   19806   6020   75843   93340   42300

3.5 Use complements to answer these problems

1. What is the complement from 100 of 67?

2. Jack buys a pen for 98 pence and a note-book for 95 pence. How much change does he receive from £2?

3. A carriage has 100 spaces and 86 of these are filled. How many spaces are left?

4. Nina uses 857 pieces of paper from her stock of 2 reams (1 ream has 500 sheets). How many pieces does she have left?

The Curious Hats of Magical Maths

2.5 Continued

5. What is 1000 minus 768?

6. Janaka has 100 cows and one day he counts 82. How many are missing?

7. How much change should be received from Rs100 when spending Rs76?

8. A theatre has 1000 seats. On one night there were 746 people in the audience. How many spare seats were there?

9. What is the difference between 672 and 1000?

10. An aeroplane has 10000km to fly. How much further has it to go after it has travelled 6765km?

11. A tall building has 100 steps. In climbing to the top a man has reached 54; how many further steps has he to go?

12. Jenna weighed out 860 grams of flour for a cake. How many grams short of 1 kilogram is this? (1000 grams = 1 kg)

13. Mary has 100 shells in her collection. If she gives 23 away to her brother Simon, how many will she have left?

14. Anish buys a book for Rs7.85. How much change will he receive from Rs10?

15. When a Roman soldier walks he makes 1000 complete paces to the mile. If a soldier walks 348 paces, how many paces has he left to complete the mile?

16. Alex buys a pen in one shop for Rs6.35 and a magnifying glass in another shop for Rs7.85. In both shops he hands over Rs10 and receives change. What is the total change from both shops?

17. Anushka finds that a quick way of multiplying 198 by 5 is to think of 200 less 2. What is her answer?

Chapter 2 Complements

Complement Puzzle

Shade pairs of squares that have a number and its complement. For example, 11 and 89 are complements so you should shade both of these squares. When you have shaded all the squares with complement pairs the unshaded squares will reveal a secret number.

81	33	62	96	72	78	58	84
57	83	14	15	60	65	93	43
67	82	32	20	97	66	99	12
3	49	40	11	45	73	53	31
80	44	2	69	92	6	91	36
75	48	34	21	63	19	39	29
88	50	41	77	7	42	10	17
5	16	54	56	64	76	25	52
55	68	23	22	87	59	38	46
27	85	8	26	71	95	50	89

Nine Times Table

The answers to the 9 times table contain 9-complements. For example, $4 \times 9 = 36$ and 3 and 6 are complements. This is used in the well-known finger system for the 9 times table.

To find 4×9 using this method hold up ten fingers and counting from the left, hold down your fourth finger. The number of fingers to the left of that one is 3 and the number to the right is 6. This makes 36.

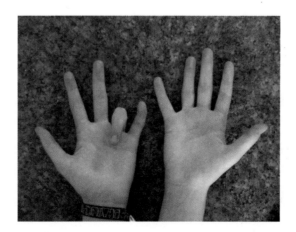

13

Chapter 3

The Deficiency

A deficiency is when there is something missing. A deficiency is a lack. 9 is deficient from 10 by 1, 8 is deficient by 2, and so on. There are many calculations where it is easier to use a deficiency than otherwise.

A deficiency is really the same as a complement but a complement is usually meant as the deficiency from 10, 100 or any other power of 10.

The thinking hat for the Deficiency is on the right.

3.1 Tuning up

1. $12 + 10 - 1 =$
2. $18 + 10 - 1 =$
3. $25 + 10 - 1 =$
4. $32 + 10 - 1 =$
5. $46 + 10 - 1 =$
6. $28 + 10 - 1 =$
7. $78 + 10 - 1 =$
8. $69 + 10 - 1 =$
9. $61 + 10 - 1 =$
10. $37 + 10 - 1 =$
11. $40 + 10 - 1 =$
12. $91 + 10 - 1 =$

Adding 9

Adding 9 is the same as adding ten and taking one off. This is very easy because you can just add 1 to the tens digit and reduce the units digit by 1.

So 26 + 9 = 35. The 2 of 26 is increased by 1 and the 6 of 26 is reduced by 1.

3.2 Adding 9

1. $27 + 9 =$
2. $38 + 9 =$
3. $67 + 9 =$
4. $84 + 9 =$
5. $28 + 9 =$
6. $33 + 9 =$
7. $56 + 9 =$
8. $88 + 9 =$
9. $21 + 9 =$
10. $73 + 9 =$
11. $243 + 9 =$
12. $456 + 9 =$
13. $213 + 9 =$
14. $376 + 9 =$
15. $553 + 9 =$
16. $7458 + 9 =$

Chapter 3 The Deficiency

By Addition and Subtraction

This sutra used for the process is *By Addition and Subtraction*.

Example 35 + 29

$$35 + 29 = 35 + 30 - 1$$
$$= 65 - 1$$
$$= 64$$

Adding 29 is the same as adding 30 and taking away 1. You can do this mentally by adding 3 to the tens digit of 35, making 6, and reducing the units digit by 1 making 4. The answer is 64.

3.3 Use the deficiency

1. 64 + 29 =
2. 37 + 29 =
3. 58 + 29 =
4. 23 + 29 =
5. 41 + 29 =
6. 35 + 29 =
7. 53 + 29 =
8. 28 + 19 =
9. 36 + 19 =
10. 55 + 19 =
11. 69 + 19 =
12. 15 + 19 =
13. 53 + 39 =
14. 26 + 49 =
15. 27 + 69 =
16. 24 + 79 =

Adding numbers close to 100 or 200, and so on, is just as easy. For example, to add 99 to a number add 1 to the hundreds digit and take 1 away from the units digit. For instance, 273 + 99 = 372.

3.4 Use the deficiency

1. 247 + 99 =
2. 762 + 99 =
3. 177 + 99 =
4. 543 + 99 =
5. 821 + 99 =
6. 116 + 99 =
7. 539 + 99 =
8. 248 + 99 =
9. 438 + 199 =
10. 566 + 299 =
11. 782 + 199 =
12. 403 + 399 =
13. 272 + 499 =
14. 246 + 399 =
15. 123 + 699 =

The Curious Hats of Magical Maths

3.5 *Here's some more practice*

1. 286 + 199 =
2. 119 + 199 =
3. 548 + 199 =
4. 387 + 199 =

5. 546 + 98 =
6. 289 + 98 =
7. 648 + 98 =
8. 539 + 98 =

9. 217 + 198 =
10. 602 + 198 =
11. 724 + 197 =
12. 645 + 298 =

Subtracting 9

Taking away 9 from a number is just as easy as adding 9. You take away 1 from the tens digit and add 1 to the units digit. For example, 57 − 9 = 57 − 10 + 1 = 48.

3.6 *Subtracting 9*

1. 27 − 9 =
2. 48 − 9 =
3. 57 − 9 =
4. 74 − 9 =
5. 41 − 9 =

6. 56 − 9 =
7. 85 − 9 =
8. 61 − 9 =
9. 73 − 9 =
10. 342 − 9 =

11. 270 − 9 =
12. 360 − 9 =
13. 537 − 9 =
14. 410 − 9 =
15. 824 − 9 =

Using a deficiency for subtraction

The examples below show how to use deficiencies for some subtractions.

46 − 19 = 27

245 − 99 = 146

678 − 199 = 479

| 19 is 1 less than 20 so take away 2 from the tens digit and add 1 to the units. | 99 is 1 less than 100 so take 100 away and add 1 to the units. | 199 is 1 less than 200 so take 2 off the hundreds digit and add 1 to the units. |

Chapter 3 The Deficiency

3.7 *Look first, then use a deficiency*

1. 45 − 19 =
2. 23 − 19 =
3. 34 − 29 =
4. 76 − 29 =
5. 435 − 99 =
6. 388 − 99 =
7. 374 − 199 =
8. 923 − 199 =
9. 84 − 49 =
10. 68 − 39 =
11. 92 − 69 =
12. 82 − 29 =
13. 200 − 99 =
14. 725 − 299 =
15. 730 − 399 =

3.8 *Here are some involving multiplication*

1. $29 \times 3 =$
2. $19 \times 2 =$
3. $39 \times 3 =$
4. $19 \times 4 =$
5. $49 \times 3 =$
6. $79 \times 2 =$
7. $199 \times 2 =$
8. $99 \times 7 =$
9. $59 \times 4 =$
10. $299 \times 3 =$
11. $28 \times 4 =$
12. $98 \times 2 =$

3.10 *Use deficiencies to work these out mentally*

1. A boy goes into a shop to buy a pen costing $3.99. How much change should he receive from $10?

2. Neda has 73 boxes of biscuits in the store of her shop. She takes out 19 boxes. How many are left in the store?

3. John has a collection of 462 stamps. He is given 29 more by his uncle. How many does he now have?

4. The book that Amy is reading has 400 pages. She has read 298 of them. How many pages does she have left to read?

5. A car park has 5 rows with 29 cars in each row. How many cars are there?

6. I am travelling 1000km from Delhi to Nagpur by train. When the train has gone 298km, how far is there to go?

Chapter 4
Number Patterns

Patterns are an important aspect of Maths. The first number pattern is simple and you will have learnt it a long time ago. It is 1, 2, 3, 4, 5, 6, 7, 8, 9, 10. The pattern continues by repeating these nine numbers together with the nought or zero. Each number is separated by 1 and each number is a one. One is the first number and you can never really get far from it.

The rule for number patterns is

> **By one more than the one before**.

Here is the magic hat.

In finding the pattern you will see that the relationship between one number and the next repeats with *One more than the one before*.

Example Find the next number in the pattern

        2   5   8   11   14

By looking at the numbers you can see that they go up by 3 each time. Each number is one more 3 than the previous. So the next number is 17.

4.1 Find the next two numbers in each pattern

1. 1 3 5 7 __ __
2. 2 4 6 8 __ __
3. 1 4 7 10 __ __
4. 5 10 15 20 __ __
5. 4 8 12 16 __ __
6. 30 33 36 39 __ __
7. 20 18 16 14 __ __
8. 7 12 17 22 __ __
9. 54 52 50 48 __ __
10. 2 8 14 20 __ __
11. 8 17 26 35 __ __
12. 2 4 8 16 __ __

Chapter 4 Number Patterns

4.2 *Here are some more for you to have a go at*

1. 2   5   8   11   ___   ___
2. 3   7   11   15   ___   ___
3. 1   2   4   7   ___   ___
4. 1   2   4   8   ___   ___
5. 0   7   14   21   ___   ___
6. 95   90   85   80   ___   ___
7. 18   27   36   45   ___   ___
8. 24   20   16   12   ___   ___
9. 22   27   32   37   ___   ___
10. 2   3   5   8   ___   ___
11. 1   3   6   10   ___   ___
12. 1   4   9   16   ___   ___

4.3 *Find the next two numbers in each pattern using the given rule*

1. 5   9   13,   add 4   ___   ___
2. 23   19,   take away 2   ___   ___
3. 34   29,   take away 5   ___   ___
4. 2   4   8,   double each number   ___   ___
5. 2   5   11,   double the number then add 1   ___   ___
6. 1   2,   multiply by 3 and take off 1   ___   ___
7. 12   23   34,   add 11   ___   ___
8. 1   4   10,   add 1 and multiply by 2   ___   ___
9. 2   3   5,   double the number and take 1 off   ___   ___
10. 2   3   6,   multiply by 3 and take 3 off   ___   ___
11. 64   32,   halve the number   ___   ___
12. 1   6   16,   add 2 then multiply by 2   ___   ___
13. 23   13,   add the digits and then add 8   ___   ___

The Curious Hats of Magical Maths

Number patterns can also appear in shapes. Look at the following pattern:

By looking at how the shape is growing you can discover further shapes in the pattern.

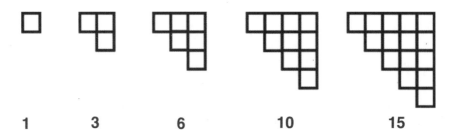

1   3   6   10   15

Remember the rule, *By one more than the one before.*

4.4 Draw the next two shapes for each pattern. Write the numbers underneath

1.

2.

3.

4.4 continued

4.

5.

6.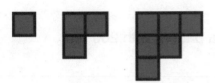

7.

8.

9.

10.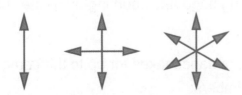

Chapter 4 Number Patterns

21

The Curious Hats of Magical Maths

4.5 Look at the sequence of patterns below and answer the questions.

1. How many red squares will there be in a pattern with 6 yellow squares?
2. How many yellow squares will there be in a pattern with 9 red squares?
3. How many red squares will there be in a pattern with 56 yellow squares?
4. How many yellow squares will there be in a pattern with 200 red squares?
5. What will be the total number of squares in a pattern with 1000 yellow squares?
6. Explain in words how you would work out the number of yellow squares if you knew the total number of squares.

4.6 Handshakes at the party!

You are invited to a party an there are 12 other people there. You shake hands with each of them. How many handshakes do you make?

Now imagine that everybody shakes hands with everybody. How many handshakes will there be all together. To investigate this, consider how many handshakes there will be with two, three, four, etc., people at the party and build up to thirteen. This can be done by joining the lettered dots on the following page and counting up all the lines the lines.

Complete the table showing the total numbers of handshakes for up to thirteen guests. The resulting numbers are called triangular numbers.

Chapter 4 Number Patterns

In each arrangement join each dot to each dot with a straight line. Use a ruler. Count the number of lines and complete the table below.

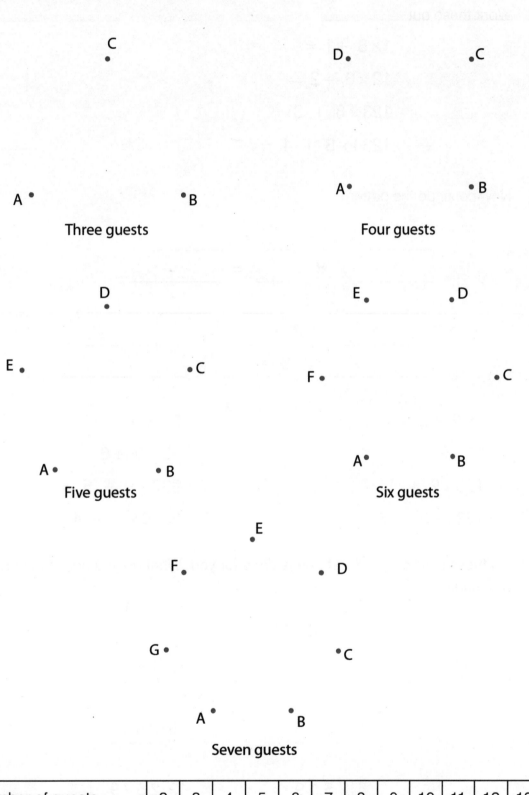

Number of guests	2	3	4	5	6	7	8	9	10	11	12	13
Number of handshakes												

The Curious Hats of Magical Maths

4.7 *Here you will find some interesting number patterns. Do the calculations on a piece of paper and then write your answers below.*

1. Work these out

    $1 \times 8 + 1 =$

    $12 \times 8 + 2 =$

    $123 \times 8 + 3 =$

    $1234 \times 8 + 4 =$

    Now continue the pattern

    _____ $\times 8 +$ ___ $=$ _____

    _____ $\times 8 +$ ___ $=$ _____

    _____ $\times 8 +$ ___ $=$ _____

    _____ $\times 8 +$ ___ $=$ _____

    _____ $\times 8 +$ ___ $=$ _____

2. $1 \times 9 + 2 =$

    $12 \times 9 + 3 =$

    $123 \times 9 + 4 =$

    $1234 \times 9 + 5 =$

3. $9 \times 9 + 7 =$

    $98 \times 9 + 6 =$

    $987 \times 9 + 5 =$

    $9876 \times 9 + 4 =$

4. Do these divisions, the first one is done for you. What do you notice about the remainders?

    $2 \overline{)25_1 1_1 9}$
    $\overline{1259/1}$

    $3 \overline{)2519}$

    $4 \overline{)2519}$

    $5 \overline{)2519}$

    $6 \overline{)2519}$

    $7 \overline{)2519}$

    $8 \overline{)2519}$

    $9 \overline{)2519}$

    $10 \overline{)2519}$

Chapter 5

Multiplication above the base

Chapter 1 dealt with multiplying numbers less than a base of 10 or 100. What happens when they are more than the base? Instead of writing down the complement you put down the surplus. For example, with 12, using 10 as the base, the surplus is 2.

This is also called *Nikhilam* multiplication because it uses the same method as before with one slight difference; you cross-add instead of cross-subtract.

Example 13×12

```
        10
     13 + 3
    ×12 + 2
    ───────
    15 / 6
```

1. Using 10 as the base, 13 is 3 more and 12 is 2 more. Put these down on the right-hand side with a connecting plus sign. The plus sign shows that 12 and 13 are both more than 10.

2. On the right-hand vertically multiply the two surpluses, 3 × 2 = 6 and write it down.

3. For the left-hand cross-add, either 13+2=15 or 12+3=15. Both give the same answer. Write that down on the left. The final answer is 156.

5.1 *Try these*

1. 11
   × 12

2. 13
   × 11

3. 11
   × 11

4. 14
   × 11

5. 15
   × 11

6. 11
   × 17

7. 12
   × 12

8. 13
   × 13

The Curious Hats of Magical Maths

What happens when the numbers are larger? The next example shows how to multiply 112 and 104. This time 100 is the base and both numbers are more than this. 100 has two zeros and so the surplus has two digits. Apart from this the method is exactly the same as before.

Example 112×104

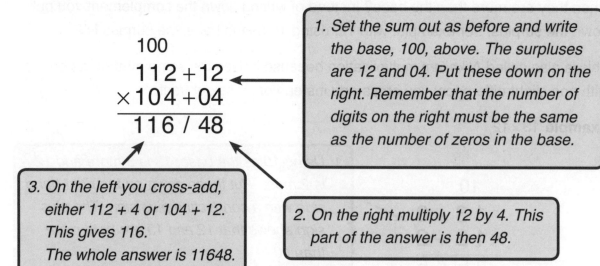

1. Set the sum out as before and write the base, 100, above. The surpluses are 12 and 04. Put these down on the right. Remember that the number of digits on the right must be the same as the number of zeros in the base.

2. On the right multiply 12 by 4. This part of the answer is then 48.

3. On the left you cross-add, either 112 + 4 or 104 + 12. This gives 116. The whole answer is 11648.

5.2 Have a go at these

1.	106 × 102	5.	108 × 101	9.	103 × 101	13.	107 × 103
2.	109 × 102	6.	107 × 102	10.	104 × 101	14.	103 × 103
3.	102 × 102	7.	106 × 101	11.	105 × 102	15.	107 × 107
4.	107 × 101	8.	108 × 102	12.	106 × 105	16.	109 × 104

Chapter 5 Multiplication above the base

5.3 *Now do these ones*

1. 108 × 103
2. 104 × 103
3. 101 × 101
4. 102 × 101
5. 103 × 102
6. 105 × 104
7. 109 × 105
8. 104 × 107
9. 104 × 104
10. 110 × 105
11. 108 × 104
12. 111 × 105
13. 106 × 103
14. 109 × 103
15. 112 × 103
16. 104 × 102

5.4 *Here are some more!*

1. 105 × 108
2. 106 × 104
3. 107 × 105
4. 105 × 103
5. 109 × 109
6. 110 × 104
7. 106 × 106
8. 105 × 101
9. 107 × 106
10. 108 × 107
11. 109 × 106
12. 110 × 103
13. 108 × 106
14. 110 × 106
15. 103 × 110
16. 112 × 106
17. 109 × 107
18. 103 × 109
19. 109 × 108
20. 112 × 108

The Curious Hats of Magical Maths

Using the diagram below you should be able to figure out how the method works. The whole rectangle measures 13 by 12 and multiplying these two gives the area. Look at the areas of the three rectangles that make up the whole.

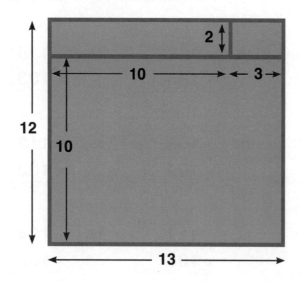
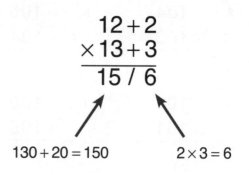

5.5 *Further practice*

1.	113 × 101	6.	111 × 103	11.	111 × 102	16.	122 × 103
2.	111 × 104	7.	112 × 108	12.	112 × 102	17.	131 × 103
3.	112 × 105	8.	132 × 102	13.	112 × 107	18.	130 × 103
4.	114 × 102	9.	115 × 102	14.	120 × 102	19.	141 × 102
5.	121 × 102	10.	107 × 107	15.	120 × 103	20.	125 × 102

Chapter 5 Multiplication above the base

Carrying to the left

Sometimes the multiplication answer on the right-hand side has more than two digits. When this happens you will have a carry digit to take over to the left. The next example shows what to do.

Example 142 × 103

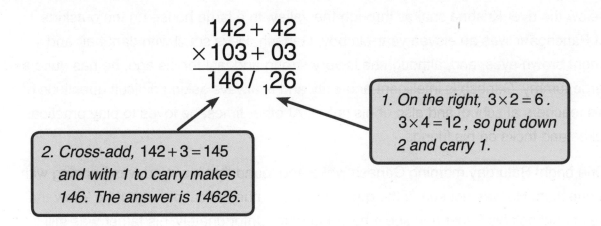

1. On the right, 3 × 2 = 6. 3 × 4 = 12, so put down 2 and carry 1.

2. Cross-add, 142 + 3 = 145 and with 1 to carry makes 146. The answer is 14626.

5.6 *With carrying*

1. 132 × 105
2. 141 × 103
3. 130 × 104
4. 123 × 106
5. 151 × 102
6. 124 × 107
7. 150 × 106
8. 162 × 102
9. 170 × 102
10. 121 × 107
11. 135 × 105
12. 153 × 102
13. 130 × 109
14. 142 × 104
15. 124 × 106
16. 152 × 102
17. 123 × 107
18. 123 × 108
19. 127 × 104
20. 163 × 102

Chapter 6

How many numbers are there?

High up in the hills of the western region of Maharashtra in India is a town called Panchgani. It is surrounded by five mountains covered with lush forests, whilst far below the river Krishna snakes through the valley. In a large house on the outskirts of Panchgani lives an eleven year-old boy, Ganesh. He is small with dark hair and bright brown eyes, and, although he is very strong and fast for his age, he has quite a large tummy. Ganesh is intelligent and witty and is always asking difficult questions of his teachers at school and also of his father. At other times, he loves to play practical jokes and tricks on his friends.

One bright Saturday morning Ganesh woke and found that his mind was bursting with a question. He was not sure if the question was mathematical or not but he knew that he should ask his father and see if he could help. Unfortunately, his father was still asleep and Ganesh thought that he must be tired and in need of extra time in bed. On the previous day his father, who was a magistrate for the court at Panchgani, had been involved with some difficult clients at his office and stayed at work until late into the night. After breakfast Ganesh sat down to do his weekend homework, an essay on climate change for Geography and an assignment on his favourite subject, Maths. Ganesh was brilliant at Maths; he loved to play with numbers in his mind and knew many properties and special relationships between them.

He was still working at the essay when his father came in, looking a little bedraggled. It was a good time to ask his question Ganesh thought.

"Good morning, Ganesh," his father said. "Have you been up for long?"

"Yes I have. I've been doing my homework. I have a question I want to ask you."

"What is it?" replied his father.

"I want to know how many numbers there are. Can you tell me this?"

His father looked at him quizzically thinking that this might be another trick question, which was a fairly common practice for Ganesh.

He decided that, although it may be a trick question, he would try to give a straight answer and even though he had only just woken up his mind was clear.

"I can think of three answers to this question. One answer is deep, another is friendly

Chapter 6 How many numbers are there?

and the third is very deep. Which would you like?"

Ganesh thought for a while before answering.

"Father, please tell me the friendly answer first".

"The friendly answer is nine," his father said wisely. "There are nine numbers and a zero. With these nine - and the zero - we can make all the numbers we want. They may repeat but all are based on these nine. It was the ancient Indians who first used numbers like this."

"So why is this the friendly answer?"

"This is the friendly answer, Ganesh, because these numbers are your friends. The best way to learn to use them is to treat them as your close friends. They are dependable and will never let you down. They are inside you and no one can take them away from you. They are also within the whole world around us and the entire creation."

"I see," said Ganesh. "So what is the deep answer?'

"The deep answer is countless," Ganesh's father replied. "Numbers go on for ever and ever, even more than the grains of sand on the sea shore. If you realise this you can touch the infinite with your imagination".

"Dad, what is the deepest answer to my question?'

"Ganesh, the deepest answer is one. There is only one number and that is the number one. If you truly understand this you will know All."

"But how can I understand this?"

"Come outside and I will show you."

Together they walked out into the dazzling sunshine of the garden and before them lay the beautiful river, widening as it flowed into the lake, sparkling in the sun. Across the valley the slopes were covered with thick green trees and rocky outcrops which grew up the sides of the hills. The air was filled with the fresh scented perfume of elegant flowers and they could hear bees buzzing at their work. From the forest they heard the morning calls of distant birds. Ganesh and his father stood for a while absorbing the breathtaking view.

"Tell me how many trees you can see?"

"There are too many, I cannot count them."

The Curious Hats of Magical Maths

"So they are countless?"

"Yes", replied Ganesh.

"Now look at this tamarind tree in our garden. How many leaves does it have?"

Ganesh looked up at the large tree indicated and considered the leaves. The thick branches were laiden with millions of small pale green leaves.

"Again, they are countless."

"Exactly so," replied his father. "But is it one tree?"

"Yes, of course."

"This is a clue to the deepest answer to your question. You can see this tree as one tree or you can see it as countless leaves. In the same way you can see anything as one or you can see anything as many.

"When you look at the tree as one you also have its name, tree. Your name is Ganesh and it stands for the person that is you. You are one person and yet you have many parts. For example, your body has feet, hands, arms, and so on. Anyone can look at you as many parts or as one person. It may take a while for you to understand all these answers and, in the meantime I have a puzzle for you.

"I draw a rectangle and divide it into quarters. The lower left-hand quarter is shaded. The upper right-hand quarter is subdivided into quarters. Of these the lower left-hand is shaded and the upper right-hand is quartered. If I were to continue like this forever what fraction of the original rectangle would be shaded?"

Ganesh's father drew the diagram on a piece of paper and handed it to Ganesh who looked at it carefully.

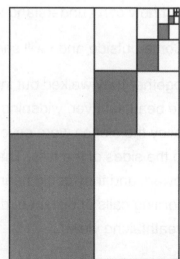

"I can see it's more than a quarter because there is more shaded than just the lower left hand rectangle. I can also see it must be less than one half but I am not sure how to solve this. I will think about it."

So he took the diagram and went to sit in his favourite spot in the garden; the den he built several weeks previously amongst some trees. He sat down and looked again at the diagram his father had drawn.

Chapter 6 How many numbers are there?

He looked at the largest shaded rectangle. It was a quarter of the whole. Then he looked at the next shaded rectangle. It was a quarter of a quarter, a sixteenth. After that he worked out that the third shaded rectangle was a quarter of a sixteenth, a sixty-fourth.

Ganesh took his pencil and under the diagram wrote down $\frac{1}{4}+\frac{1}{16}+\frac{1}{64}+\cdots=$

Now he could see that he would have to add up an infinite number of fractions each one four times smaller than the last. He looked up at the tamarind tree and wondered at the impossibility of counting the leaves and pondered on how to do this in the puzzle. He just could not see how to do it. Ganesh decided to start adding the fractions to see if there was a pattern.

He added the first two fractions together, $\quad \frac{1}{4}+\frac{1}{16} = \frac{4}{16}+\frac{1}{16} = \frac{5}{16}$

He then added the next fraction, $\quad \frac{1}{4}+\frac{1}{16}+\frac{1}{64} = \frac{16}{64}+\frac{4}{64}+\frac{1}{64} = \frac{21}{64}$

Adding the next fraction took him a while,

$$\frac{1}{4}+\frac{1}{16}+\frac{1}{64}+\frac{1}{256} = \frac{64}{256}+\frac{16}{256}+\frac{4}{256}+\frac{1}{256} = \frac{85}{256}$$

He then stopped and looked at his answers. Was there a pattern? He looked at the top numbers, 5, 21, 85 and after a while saw that each number could be got from the one before by multiplying by 4 and adding 1.

$$1\times 4+1=5, \quad 5\times 4+1=21, \quad 21\times 4+1=85$$

He also saw that the top number of each fraction, when multiplied by 3, was one less than the bottom number. So he wrote down this,

$$5\times 3+1=16, \quad 21\times 3+1=64, \quad 85\times 3+1=256$$

But it was no good. He tried and tried but could not see how to get to the answer, the numbers were just getting too large. And so at last he decided to give up. He went in search for his father to see if he would help and found him in his study immersed in legal papers. Ganesh waited for his father's attention and then showed him his workings.

"This is how far I've got but I cannot seem to get any further. Please can you give me a clue."

The Curious Hats of Magical Maths

Ganesh's father looked at the calculations and after a short pause said, "I will give you two clues. The first clue is the mathematical rule, *As in one, so in all*. You must think of the individual and you must think of the whole. The second clue is to cut the rectangle up into "L" shapes."

Ganesh thanked his father and went into the kitchen for a pair of scissors. He started to cut out the "L" shapes and laid them out on a table. He could not cut out more than six as they were getting very small. Nevertheless, he spread them out neatly and looked carefully. Suddenly, a short flash of inspiration came to him as bright as the day around him. In an instant he realised the answer and saw why he was told the rule and also why he should cut out the "L" shapes.

The answer was one third!

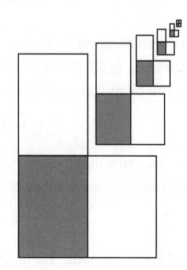

Each piece has one third part shaded and since each is the same shape as all the others then one third of the whole is shaded.

Ganesh was very pleased with this problem because the solution was so easy when you knew how to get there. He also felt happier about understanding his father's answers but these things he would need to consider further. In the meantime, he would try this puzzle out on his friends.

Chapter 7

Multiplication by *Vertically and Crosswise*

Multiplication by the *Vertically and Crosswise* rule works for multiplying any two numbers together in one line. In this chapter you will see how to multiply two-digit numbers together.

Example 21×32

```
    2 1
  × 3 2
  ─────
  6 7 2
```

1. Set out the sum with one number below the other. Leave a space between the two digits of each number.

2. Multiply vertically $1 \times 2 = 2$ and set this down on the right.

3. The middle step is to multiply crosswise $2 \times 2 = 4$, and $3 \times 1 = 3$, and then add these two results making 7. This is placed in the space to the left of 2.

4. Finally, multiply vertically, $3 \times 2 = 6$, and set this down on the left.
The answer is 672.

7.1 *Follow the steps shown above*

1.   1 3
   × 1 2

2.   2 1
   × 3 1

3.   2 2
   × 3 1

4.   3 2
   × 2 1

5.   4 0
   × 2 1

6.   2 2
   × 3 0

7.   2 3
   × 2 1

8.   2 2
   × 2 2

9.   1 1
   × 1 2

10. 3 1
   × 3 1

11. 1 4
   × 1 1

12. 1 3
   × 1 1

The Curious Hats of Magical Maths

This next example shows carrying.

Example 42×13

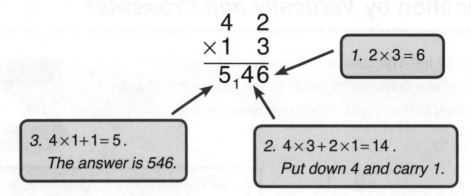

1. $2 \times 3 = 6$
2. $4 \times 3 + 2 \times 1 = 14$. Put down 4 and carry 1.
3. $4 \times 1 + 1 = 5$. The answer is 546.

7.2 With carrying

1. $\begin{array}{r} 24 \\ \times 13 \\ \hline \end{array}$
2. $\begin{array}{r} 12 \\ \times 16 \\ \hline \end{array}$
3. $\begin{array}{r} 34 \\ \times 21 \\ \hline \end{array}$

4. $\begin{array}{r} 25 \\ \times 13 \\ \hline \end{array}$
5. $\begin{array}{r} 26 \\ \times 21 \\ \hline \end{array}$
6. $\begin{array}{r} 31 \\ \times 41 \\ \hline \end{array}$

7. $\begin{array}{r} 22 \\ \times 32 \\ \hline \end{array}$
8. $\begin{array}{r} 43 \\ \times 12 \\ \hline \end{array}$
9. $\begin{array}{r} 42 \\ \times 41 \\ \hline \end{array}$

10. $\begin{array}{r} 52 \\ \times 23 \\ \hline \end{array}$
11. $\begin{array}{r} 17 \\ \times 18 \\ \hline \end{array}$
12. $\begin{array}{r} 22 \\ \times 19 \\ \hline \end{array}$

The area problem below shows you how the method works for $21 \times 32 = 672$. The whole rectangle, not to scale, measures 32 by 21 an its area is 672. This is made of up four rectangles, A is $2 \times 1 = 2$ which is the first step in the calculation.

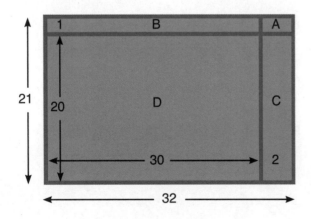

B is $30 \times 1 = 30$ and C is $2 \times 20 = 40$ and these taken together make 70. This is the middle step.

D is $30 \times 20 = 600$ which is the last step.

$\begin{array}{r} 21 \\ \times 32 \\ \hline 672 \end{array}$

Chapter 7 Multiplication by Vertically and Crosswise

7.3 Further practice

1. 22 × 32
2. 35 × 12
3. 32 × 13
4. 14 × 39
5. 32 × 33

6. 17 × 29
7. 27 × 14
8. 16 × 32
9. 76 × 11
10. 14 × 23

11. 35 × 22
12. 26 × 14
13. 41 × 51
14. 15 × 45
15. 13 × 19

16. 49 × 15
17. 16 × 53
18. 12 × 48
19. 59 × 17
20. 26 × 24

7.4 Set out each problem in the box on the right.

1. Multiply 46 by 32.

2. Find the product of 23 and 48.

3. What is 53 times 84?

The Curious Hats of Magical Maths

7.4 continued

4. Multiply forty-two by twenty-eight.

5. Find the cost of sixteen radios at £53 each.

6. If there are 27 girls in a class and each one has 14 crayons, how many crayons are there altogether.

7. What is twenty-four lots of 12?

8. Find the product of thirty-eight and sixteen.

9. A coach company has 21 coaches and each coach can carry 53 passengers. How many passengers can all the coaches carry?

10. A block of stamps has 24 rows with 14 in each row. How many stamps are there in the block?

11. If a packet of biscuits costs 64p, find the cost of a whole box containing forty-eight packets.

Chapter 7 Multiplication by Vertically and Crosswise

7.5 *Here are some more problems. Work them out and then draw a line from the box to the correct answer on the right.*

1. Multiply 62 by 71.

2. Find the product of 34 and 18.

    744

    384

3. A cinema has 16 rows with 24 seats in each row. How many seats are there altogether?

    4402

    612

4. Calculate the number of hours in the month of January.

    336

    350

5. A girl learnt 20 verses of scripture a day for each of 48 days. How many verses is this?

    2016

    350

6. If you can do twenty-five sums a day, how many sums can you do in fourteen days?

    960

7. A bricklayer builds a wall with 72 bricks in each of 28 rows. How many bricks does he use altogether?

8. A pastry chef bakes tarts in 28 trays each containing 12 tarts. How many tarts does the chef bake?

39

The Curious Hats of Magical Maths

Three digit multiplication

The *Vertically and Crosswise* rule gives a universal method for multiplying numbers of any size in one line. The next example shows how to multiply two three digit numbers together. The Vertically and Crosswise rule gives a universal method for multiplying numbers of any size in one line.

The pattern of steps for three by three digit numbers is shown on the right. The six dots represent the digits of each number.

Example 231×124

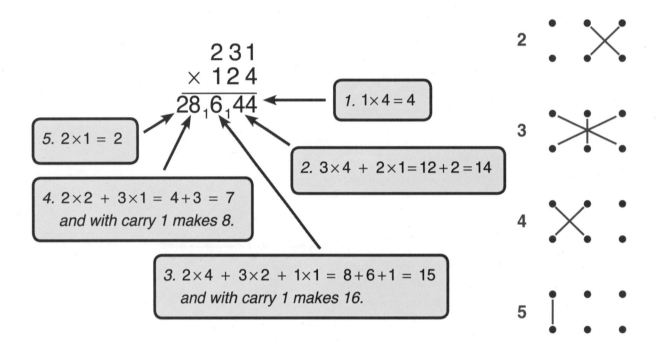

7.6 Follow the steps and answer these

1.	3 2 1 × 2 1 1	4.	1 2 2 × 2 1 1	7.	7 2 0 × 1 1 2	10.	1 6 4 × 3 2 1
2.	2 3 1 × 1 2 4	5.	5 3 2 × 1 1 1	8.	3 4 0 × 2 4 1	11.	1 0 7 × 5 2 0
3.	3 0 2 × 1 3 4	6.	2 2 3 × 3 0 1	9.	2 5 1 × 3 4 2	12.	4 3 9 × 2 1 2

7.7 These are a little harder

1. 547
    × 823

2. 659
    × 335

3. 378
    × 473

4. 869
    × 374

5. 578
    × 465

6. 482
    × 536

7. 864
    × 753

8. 309
    × 495

9. 699
    × 712

To multiply a two-digit number and a three-digit number together you can use the same steps but place a 0 to the left of the two-digit number, as if it has three-digits.

Example 23 × 324

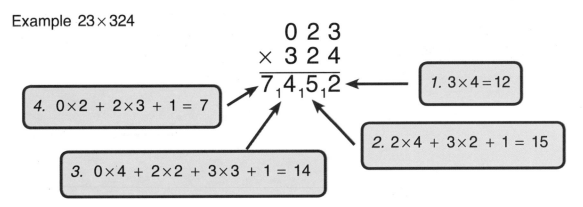

7.8 There is one less step in these

1. 34
    × 225

2. 52
    × 304

3. 511
    × 42

4. 620
    × 53

5. 42
    × 371

6. 28
    × 531

7. 31
    × 712

8. 24
    × 235

9. 825
    × 34

41

Chapter 8

Subtraction by *On The Flag*

This easy method for subtracting numbers uses the *On the flag sutra*.

The method works from the left in the same way that division is from the left.

Example 72 − 48

1. Starting on the left, 7 minus 4 is 3. Look ahead to the next column, 8 is more than 2 and so cannot be subtracted. Take one away from 3 leaving 2 and write this in the first column.

2. You have taken 1 away from the 3 in the first column and you have to give it back as 10 in the last column. Place the 1 that you've taken away On the flag, to the left of the 2. This now makes 12.

3. In the right hand column, 12 minus 8 leaves 4. The answer is then 24.

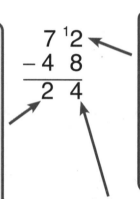

8.1 *Start from the left and remember to look ahead*

1.	53 − 29	4.	55 − 26	7.	71 − 28	10.	67 − 39
2.	61 − 36	5.	75 − 37	8.	92 − 38	11.	66 − 37
3.	72 − 58	6.	81 − 43	9.	95 − 48	12.	76 − 58

Chapter 8 Subtraction by On the Flag

Example 764 − 587

7^16^14
$-\ 5\ 8\ 7$
$\overline{1\ 7\ 7}$

1. Starting from the left, 7 minus 5 is 2. Look ahead, 8 is bigger than 6, so take 1 away from the 2, put down 1 and put one as a flag digit behind the 6.

2. 16 minus 8 leaves 8. Look ahead to the next column, 7 is bigger than 4, so take 1 away from 8 leaving 7 and put a 1 as a flag digit in the next column.

3. In the last column, 14 minus 7 is 7. The answer is 177.

8.2 Subtracting three digit numbers

1. 804
    − 257

2. 725
    − 386

3. 532
    − 297

4. 900
    − 432

5. 712
    − 256

6. 750
    − 276

7. 430
    − 271

8. 752
    − 594

9. 633
    − 587

10. 858
    − 489

11. 550
    − 398

12. 682
    − 295

When digits in the following column are the same you have to look ahead one further column.

Example 734 − 238

7^13^14
$-\ 2\ 3\ 8$
$\overline{4\ 9\ 6}$

1. Starting from the left, 7 minus 2 is 5. Look ahead, 3 is the same as 3, so look ahead one more column, 8 is bigger than 4. Take 1 off the 5, leaving 4, and place a 1 in the flag position.

2. 13 minus 3 leaves 10. Look ahead to the next column, 8 is bigger than 4, so take 1 away from 10 leaving 9 and place the flag digit next to the 4 making it 14.

3. 14 minus 8 is 6. The answer is 496.

The Curious Hats of Magical Maths

8.3 *Now try these*

1. 842
    −247

2. 573
    −279

3. 822
    −523

4. 985
    −488

5. 872
    −275

6. 435
    −139

7. 802
    −507

8. 763
    −466

9. 494
    −395

10. 702
    −108

11. 855
    −558

12. 421
    −323

13. 913
    −319

14. 783
    −585

15. 691
    −298

16. 777
    −272

You can now try larger subtractions. Follow the steps in the next example.

Example 7984 − 2925

$$\begin{array}{r} 798^14 \\ -2925 \\ \hline 5059 \end{array}$$

1. 7 minus 2 is 5. Look ahead, 9 is the same as 9, so look ahead one more column, 8 is bigger than 2, this is ok. Put down 5 as the first answer digit. 9 minus 9 is 0.

2. 8 minus 2 is 6, look ahead, 5 is bigger, so take 1 away from 6 and place the flag digit, making 14 in the top.

3. 14 minus 5 is 9. The answer is 5059.

8.4 *Further practice*

1. 5463
    −3438

2. 7861
    −2564

3. 9832
    −4595

4. 5920
    −2496

Chapter 8 Subtraction by On the Flag

8.4 Continued

5. 7721
    −3056

6. 7541
    −2091

7. 9574
    −2275

8. 3536
    −2739

9. 8760
    −3281

10. 8560
    −2535

11. 7324
    −4359

12. 9371
    −6678

13. 5932
    −4967

14. 8484
    −5937

15. 4173
    −2325

16. 5272
    −1275

17. 9166
    −3869

18. 7020
    −5026

19. 8053
    −1058

20. 8324
    −7327

8.5 Set out each problem in the box on the right.

1. Find the difference between 763 km and 489 km.

2. Find the difference between the heights of Niki and Sara if Niki is 167 cm tall and Sara is 129 cm tall.

3. A newspaper seller has 850 copies of today's paper. How many has he left after he has sold 498?

4. A farmer has 253 lambs and sells 196 of them at the market. How many does he have left?

Chapter 9

Nikhilam Subtraction

This chapter is called Nikhilam subtraction because it uses the *All from nine and the last from ten* rule to do subtractions. The rule, or sutra, was first written in the Sanskrit language and starts with the word *Nikhilam*, hence the name. The method described here is completely different from that of the previous chapter and you may like to have a go and then compare the two methods. Vedic maths often has more than one way of solving a problem. The idea is that you can choose the one that suits you best or which gets to the answer most easily.

The method uses complements from 9 and complements from 10.

Example 52 − 29

1. 9 is bigger than 2 and so cannot be subtracted. To deal with this take the difference between 9 and 2, that is 7, and write down its ten-complement, 3.

2. In the tens column find the difference between 2 and 5, that is 3, and subtract 1 leaving 2. The answer is 23.

Subtracting 1 in the last step takes you out of complements.

9.1

1.	21 −17	4.	22 −19	7.	42 −33	10.	54 −15
2.	42 −35	5.	35 −29	8.	76 −48	11.	50 −16
3.	64 −26	6.	46 −37	9.	62 −34	12.	64 −55

9.2 Further practice

1. 81 − 12
2. 11 − 9
3. 47 − 39
4. 65 − 37
5. 72 − 56
6. 84 − 17
7. 17 − 9
8. 37 − 28
9. 80 − 37
10. 36 − 28
11. 51 − 24
12. 30 − 28
13. 82 − 19
14. 26 − 19
15. 50 − 26
16. 47 − 39

Three digit subtraction

When working through a subtraction, once the 10-complement has been used you use 9-complements.

Example 627 − 259

```
  627
− 259
  ───
  368
```

1. On seeing that 9 is more than 7 the first step is to find the difference, 2, and then take that away from 10, giving 8.

2. In the tens column, 5 is more than 2. Find the difference, 3, and subtract that from 9, giving 6.

3. In the hundreds column, 6 is more than 2. Find the difference, 4, and subtract 1 leaving 3. The answer is 368.

47

To go into complements the first is from 10. After that you just use the 9-complement until the top digit is larger than the bottom digit and you don't need complements.

9.3

1. 423
   −134

2. 762
   −673

3. 581
   −192

4. 721
   −345

5. 536
   −247

6. 565
   −177

7. 405
   −186

8. 254
   −168

9. 283
   −194

10. 960
    −471

11. 626
    −238

12. 555
    −366

9.4 Further practice

1. 367
   −298

2. 246
   −157

3. 600
   −348

4. 911
   −582

5. 776
   −689

6. 617
   −128

7. 888
   −699

8. 724
   −456

9. 305
   −116

10. 342
    −156

11. 743
    −357

12. 562
    −076

13. 376
    −297

14. 241
    −153

15. 354
    −276

16. 500
    −379

Chapter 9 Nikhilam Subtraction

Starting with complements in the middle

Example 738 − 452

```
  738
− 452
  286
```

1. In the units column, 8 − 2 = 6.

2. In the tens column 5 is more than 3 and so complements are needed. Find the difference, 2, and write down the ten-complement, 8

3. In the hundreds column, 7 is more than 4. Find the difference, 3, and subtract 1 leaving 2. The answer is 286.

9.5 Begin complements in the tens column

1.	436 − 181	7.	510 − 360	13.	467 − 393	19.	277 − 185
2.	614 − 154	8.	546 − 454	14.	824 − 531	20.	732 − 641
3.	720 − 650	9.	327 − 042	15.	345 − 260	21.	348 − 172
4.	853 − 272	10.	305 − 183	16.	275 − 191	22.	647 − 556
5.	806 − 444	11.	269 − 187	17.	462 − 380	23.	873 − 781
6.	727 − 277	12.	246 − 152	18.	211 − 190	24.	562 − 392

The Curious Hats of Magical Maths

9.6 *Set each subtraction problem out in the box on the right.*

1. A mother spends Rs874 on shopping for food. How much chamge does she have from Rs1000?

2. A hospital has 438 beds. How many beds are available when there are 287 patients already?

3. A charity tries to find orphanage places for 724 children but can only find 498 places. How many children are left without a place?

4. A theatre can seat 740. One night there were 136 spare seats. How many people were in the audience?

5. At the fair Rajvi guessed the number of sweets in a jar as 386. The correct number was 435. By how many was he wrong?

6. A boat trip to an island takes 245 minutes. How long is there after 187 minutes?

7. At a certain school there are 184 places for new pupils. If 321 apply, how many do not obtain a place?

8. On a 720 km train journey I have gone 358 km. How far have I left to go?

Chapter 9 Nikhilam Subtraction

In and out of complements

When working through a subtraction sum from the right go *into* complements when the bottom digit is bigger than the digit above and only come *out* of complements when the top digit is bigger than the bottom digit.

Example 768435 − 372837

```
  768435
− 372837
  ‾‾‾‾‾‾
  395598
```

6. 7 minus 3 is 4 and drop 1 leaves 3. The answer is 395598.

1. 7 is bigger than 5. The difference is 2, 10-complement, 8.

5. 7 is bigger than 6 so go into complements here. The difference is 1 and the 10-complement is 9.

2. Stay in complements until the top digit is larger. Difference, 0, 9-complement is 9.

4. 8 is bigger than 2 so come out of complements. The difference is 6, drop 1 leaves 5.

3. 8 is bigger than 4 so stay in complements. The difference is 4, 9-complement, 5.

9.7 *In and out of complements. See if you can do this exercise in 3 minutes!*

1.	65738 − 27293	5.	7546390 − 6748339	9.	75693002 − 16873295
2.	944751 − 638926	6.	4100288 − 2700819	10.	64538473 − 25649585
3.	800463 − 503199	7.	7000000 − 3647532	11.	74633390 − 42709983
4.	745893 − 489748	8.	9546352 − 3546372	12.	35467253 − 15577357

The Curious Hats of Magical Maths

Find the hidden number

9.8 Answer these and then find them in the grid. Shade in the boxes with the answers in to reveal a hidden number.

1. 8206 − 3557	5. 5134 − 1765	9. 7630 − 3683	13. 8283 − 5919
2. 5446 − 2898	6. 7500 − 3259	10. 5722 − 4697	14. 5372 − 2417
3. 7120 − 4396	7. 4279 − 3686	11. 8279 − 4092	15. 9283 − 4956
4. 3000 − 2643	8. 8320 − 3180	12. 7372 − 2737	16. 5472 − 3865

4378	2724	593	4649	1607	2646
534	9453	3215	7004	2548	2702
3245	8459	1722	3035	4187	1956
4231	4241	4635	357	4327	1133
7558	3055	6811	1719	5140	5634
1947	4558	2310	768	3369	2581
5227	1025	3947	2955	2364	622

Chapter 10

Nikhilam Division

Division by 9

There is a very easy way to divide numbers by 9 that doesn't look like division at all!

Example 121÷9

```
9 | 1 2 / 1
  |   1 3
  ———————
    1 3 / 4
```

1. Set the sum out as shown. Place a remainder stroke one digit in from the right-hand end.

2. The first step is to bring the first digit, 1, down into the answer line. You then write this under the next digit, 2, and add up the second column. 2 + 1 = 3, and this is the next answer digit.

3. Again, write this answer digit in the next column and add up for the remainder, 4. The answer is 13 remainder 4.

In fact you are finding the answer just by adding the digits. The reason this works is because 9 is 1 less than 10. Have a look at the very simple examples below to see if you can understand why the method works.

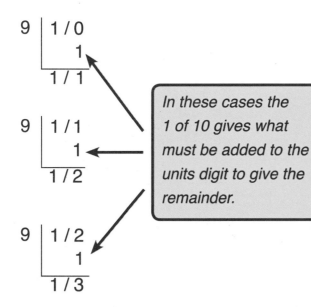

In these cases the 1 of 10 gives what must be added to the units digit to give the remainder.

```
9 | 2 / 1
  |     2
  ———————
    1 / 3
```

With 21 ÷ 9 above there is 1 for each 10 to be added to the units digit. 20 is 2 tens and so 2 is added to 1 to give the remainder 3.

The Curious Hats of Magical Maths

10.1 First steps

1. 9) 3/2
2. 9) 2/4
3. 9) 4/2
4. 9) 6/1
5. 9) 5/3
6. 9) 3/3
7. 9) 1/5
8. 9) 8/0
9. 9) 7/0
10. 9) 5/0
11. 9) 3/0
12. 9) 2/5
13. 9) 4/3
14. 9) 2/2
15. 9) 5/2
16. 9) 7/1
17. 9) 4/0
18. 9) 2/6
19. 9) 3/1
20. 9) 4/4
21. 9) 5/1
22. 9) 6/2
23. 9) 1/7
24. 9) 3/5

10.2 These are also very easy

1. 9) 10/3
2.) 20/4
3. 9) 30/0
4. 9) 11/2
5. 9) 10/4
6. 9) 23/2
7. 9) 14/2
8. 9) 12/3

Chapter 10 Nikhilam Division

10.2 continued

9. 9 | 1 0 / 7

10. 9 | 1 4 / 3

11. 9 | 2 0 / 1

12. 9 | 2 1 / 1

13. 9 | 3 2 / 1

14. 9 | 4 1 / 2

15. 9 | 5 0 / 3

16. 9 | 6 2 / 0

17. 9 | 6 1 / 1

18. 9 | 2 3 / 2

19. 9 | 5 3 / 0

20. 9 | 4 2 / 2

21. 9 | 2 5 / 1

22. 9 | 4 2 / 1

23. 9 | 1 6 / 1

24. 9 | 5 2 / 1

10.3 These are longer but the method is exactly the same

1. 9 | 1 1 2 / 1

2. 9 | 1 2 4 / 1

3. 9 | 2 1 1 / 1

4. 9 | 3 1 2 / 1

5. 9 | 3 2 1 / 1

6. 9 | 2 3 2 / 0

7. 9 | 3 2 3 / 0

8. 9 | 2 0 0 / 0

9. 9 | 4 1 0 / 2

10. 9 | 2 0 0 / 2

11. 9 | 3 2 1 / 0

12. 9 | 2 1 0 / 5

13. 9 | 1 3 2 / 2

14. 9 | 4 2 0 / 0

15. 9 | 3 1 0 / 4

16. 9 | 3 0 0 / 0

The Curious Hats of Magical Maths

What to do when the remainder is too large

By adding the digits you can see that sometimes the remainder will by 9 or more than 9. When this happens you need to divide that remainder by 9.

Example 29 ÷ 9

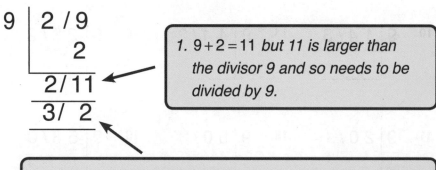

1. 9 + 2 = 11 *but 11 is larger than the divisor 9 and so needs to be divided by 9.*

2. 9 ÷ 2 *is 1 remainder 2. The 1 is added to the quotient digit 2, making 3 and the answer is then 3 remainder 2.*

10.4 Dividing the remainder

1. 9 | 4 / 6
2. 9 | 5 / 4
3. 9 | 8 / 2
4. 9 | 2 2 5
5. 9 | 1 3 6
6. 9 | 2 3 7
7. 9 | 1 6 2
8. 9 | 1 8 4
9. 9 | 2 2 6
10. 9 | 4 1 7
11. 9 | 3 1 6
12. 9 | 2 6 2
13. 9 | 4 3 3
14. 9 | 6 2 4
15. 9 | 7 1 8
16. 9 | 8 1 9

Chapter 10 Nikhilam Division

Dividing using the complement

So far it has not been obvious why this is called *Nikhilam* division and why the *All from 9 and the last from 10* rule is involved. The reason is that the complement of the divisor is used but when dividing by 9, which has a complement of 1, there is no effect.

Example 111÷8

1. Set the sum out as shown and write the complement, 2, under the divisor, 8. Bring down the first dividend digit is brought down in the answer line.

2. Multiply the answer digit, 1, by the complement, 1×2=2, and write this in the next column, to the right, underneath the 1.

3. Add up the second column, 1 + 2 = 3, and write this as the next answer digit.

4. Multiply the second answer digit, 3 by the complement, 3×2=6, and put this in the next column to the right underneath the 2.

5. Add up for the remainder, 1 + 6 = 7. The answer is then 13 remainder 7.

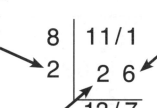

10.5 Nikhilam division with single digit divisors

1. 8 | 1 0
2. 8 | 1 0 0
3. 8 | 2 0
4. 8 | 1 1

5. 8 | 1 5
6. 7 | 1 2
7. 6 | 1 1
8. 8 | 1 3

9. 8 | 2 2
10. 7 | 1 1
11. 6 | 1 0
12. 8 | 2 3

13. 8 | 3 1
14. 8 | 1 0 2
15. 8 | 3 0
16. 8 | 1 0 1

57

The Curious Hats of Magical Maths

Larger Divisors

When dividing by a number close to the base 100, but less than 100, like 93 or 87, you can use *All from 9 and the last from 10* to obtain the deficiency, or complement as it is called

Example 2103 ÷ 88

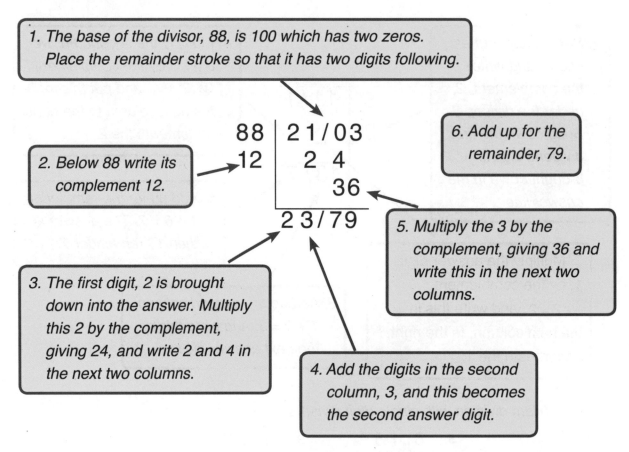

1. The base of the divisor, 88, is 100 which has two zeros. Place the remainder stroke so that it has two digits following.
2. Below 88 write its complement 12.
3. The first digit, 2 is brought down into the answer. Multiply this 2 by the complement, giving 24, and write 2 and 4 in the next two columns.
4. Add the digits in the second column, 3, and this becomes the second answer digit.
5. Multiply the 3 by the complement, giving 36 and write this in the next two columns.
6. Add up for the remainder, 79.

The number of digits to the left of the remainder stroke in the dividend tells you how many rows you will need. In the example below three rows are needed.

```
98 | 500/00
02 |  1 0
   |    0 2
   |       0 0
   |————————
     5 1 0/2 0
```

What's a dividend? Here's a rhyme to help you remember the parts of a division:

The divisor is the number that divides the dividend,

the answer is the quotient, the remainders at the end.

Chapter 10 Nikhilam Division

10.6 *Nikhilam division with double digit divisors, base 100*

1. 99 | 2 0 0
2. 98 | 2 0 0 0
3. 88 | 2 0 0 0
4. 89 | 2 0 0 0 0
5. 92 | 3 0 0 0
6. 98 | 3 0 0 0

10.7 *Check your answers to the previous exercise and then do these*

1. 99 | 2 0 0 0 0
2. 89 | 2 0 0
3. 98 | 2 0 0 0 0
4. 88 | 1 0 0 0 0
5. 97 | 1 0 0 0 0
6. 99 | 9 0 0 0 0
7. 87 | 1 0 0 0 0
8. 79 | 1 0 0 0
9. 88 | 1 0 0 1 3
10. 86 | 1 0 0 0
11. 95 | 2 0 0 0 0
12. 89 | 3 0 0 0
13. 87 | 2 0 0 0
14. 99 | 5 0 0 0 0
15. 78 | 1 0 1 3
16. 96 | 5 0 1 0 0
17. 94 | 2 0 0 0 0
18. 91 | 9 0 0 0

59

The Curious Hats of Magical Maths

10.8

1. 98) 142	10. 94) 841	19. 97) 481
2. 99) 243	11. 93) 547	20. 91) 815
3. 94) 403	12. 74) 221	21. 95) 830
4. 98) 604	13. 89) 425	22. 85) 419
5. 59) 105	14. 76) 131	23. 94) 341
6. 75) 213	15. 82) 242	24. 69) 201
7. 88) 512	16. 80) 319	25. 73) 212
8. 72) 141	17. 87) 408	26. 81) 211
9. 79) 306	18. 81) 231	27. 89) 711

10.9

1. 80 | 3 1 5

2. 69 | 2 0 5

3. 85 | 1 4 9

4. 95 | 2 7 7

5. 99 | 9 8 0

6. 91 | 7 0 3

7. 89 | 6 1 3

8. 84 | 2 4 6

9. 60 | 1 1 6

10. 99 | 2 0 3

11. 98 | 1 1 0 3

12. 97 | 2 0 1 7

13. 98 | 1 2 1 2 1

14. 89 | 1 1 0 1 1

15. 87 | 2 1 1

16. 88 | 1 0 2 0 2

17. 95 | 2 0 3 1 2

18. 77 | 1 0 0 0

19. 94 | 2 0 1 0 3

20. 83 | 1 3 5

21. 93 | 2 2 1 5

22. 76 | 2 1 2

23. 92 | 3 2 3

24. 96 | 2 3 0 3

25. 87 | 1 0 2 8

26. 95 | 1 0 0 5

27. 84 | 2 0 2 0

Chapter 10 Nikhilam Division

The Curious Hats of Magical Maths

This method can easily be extended to very large divisors.

10.10 Base 1000. The first one is done for you.

1. 889 | 10/203
   111 | 1 1 1
        | 1 1 1
   ─────────────
        | 11/ 424

2. 888 | 10/203

3. 799 | 10/203

4. 968 | 11204

5. 857 | 11013

6. 936 | 21037

7. 918 | 41230

8. 824 | 3256

9. 742 | 1356

10. 998 | 21410

11. 928 | 22146

12. 982 | 20468

13. 928 | 10145

14. 874 | 11235

15. 956 | 11025

16. 946 | 21246

17. 772 | 2142

18. 987 | 21034

You must remember this is a special case method and should only be used when the divisor is a little less than a power of ten. For divisors more than a power of ten, *Paravartya* division is used and for general division, *Straight* division, using *On the Flag*, is used. These are taken up in Book 2.

Chapter 11

Proportionately

The *Proportionately* rule is used for some aspects of multiplication and division. It is also used for reducing fractions to lowest terms and for problems with ratios.

Halving and doubling

11.1 Halve each of these numbers by dividing by 2

1. 16 _____
2. 20 _____
3. 30 _____
4. 24 _____
5. 18 _____
6. 40 _____
7. 28 _____
8. 36 _____
9. 50 _____
10. 32 _____
11. 48 _____
12. 64 _____
13. 82 _____
14. 56 _____
15. 72 _____
16. 92 _____

11.2 Halve these

1. 60 _____
2. 120 _____
3. 240 _____
4. 160 _____
5. 420 _____
6. 820 _____
7. 150 _____
8. 500 _____
9. 110 _____
10. 230 _____
11. 570 _____
12. 830 _____
13. 112 _____
14. 376 _____
15. 838 _____
16. 914 _____

11.3 Double each of these numbers by multiplying by 2

1. 14 _____
2. 21 _____
3. 34 _____
4. 23 _____
5. 40 _____
6. 15 _____
7. 47 _____
8. 53 _____
9. 11 _____
10. 72 _____
11. 35 _____
12. 45 _____
13. 56 _____
14. 74 _____
15. 89 _____
16. 65 _____

Multiplying and dividing by 5

To multiply a number by 5 first multiply it by 10 and then halve your answer.

Example 8×5

$$8 \times 5 = 8 \times 10 \div 2 = 80 \div 2 = 40$$

Alternatively you can divide by 2 first and then multiply by 10.

$$8 \times 5 = 8 \div 2 \times 10 = 4 \times 10 = 40$$

11.4 *Mentally!*

1. $9 \times 5 =$
2. $14 \times 5 =$
3. $16 \times 5 =$
4. $22 \times 5 =$
5. $18 \times 5 =$
6. $32 \times 5 =$
7. $30 \times 5 =$
8. $24 \times 5 =$
9. $50 \times 5 =$
10. $84 \times 5 =$
11. $44 \times 5 =$
12. $36 \times 5 =$
13. $48 \times 5 =$
14. $38 \times 5 =$
15. $52 \times 5 =$

Dividing is just as easy. You just have to reverse the process so divide by 10 and multiply by 2, or vice versa.

Example $120 \div 5$

$$120 \div 5 = 120 \div 10 \times 2 = 12 \times 2 = 24$$
$$\text{or}$$
$$120 \div 5 = 120 \times 2 \div 10 = 240 \div 10 = 24$$

11.5 Mentally!

1. $20 \div 5 =$
2. $80 \div 5 =$
3. $140 \div 5 =$
4. $180 \div 5 =$
5. $300 \div 5 =$
6. $70 \div 5 =$
7. $100 \div 5 =$
8. $150 \div 5 =$
9. $190 \div 5 =$
10. $410 \div 5 =$
11. $630 \div 5 =$
12. $250 \div 5 =$
13. $320 \div 5 =$
14. $620 \div 5 =$
15. $430 \div 5 =$

Multiplying and dividing by 4

One way of multiplying by 4 is to double and double again. Similarly you can divide a number by 4 by halving and halving again.

Example 13×4

$$13 \times 4 = 13 \times 2 \times 2 = 26 \times 2 = 52$$

Example $68 \div 4$

$$68 \div 4 = 68 \div 2 \div 2 = 34 \div 2 = 17$$

The Curious Hats of Magical Maths

11.6 Do these in two steps

1. $62m \times 4$
2. $100 yrs \div 4$
3. 125×4
4. $Rs 3.11 \times 4$
5. $84 cm \div 4$
6. $64 kg \div 4$
7. 121×4
8. $480 m \div 4$

9. £$248 \div 4$
10. $19 km \times 4$
11. $37 kg \times 4$
12. $128 cm \div 4$
13. $\$135 \times 4$
14. $684 m \div 4$
15. $Rs 500 \div 4$
16. £231×4

11.7 Easy proportion problems. Write your answers in the boxes.

1. Find the weight of 6 boxes of oranges if each box weighs 11 kg.
2. If a shoe box contains 2 shoes how many shoes are there in 8 boxes?
3. A pack of 4 bottles of water costs $6. How much will 12 bottles cost?
4. A motorbike can travel 30km on 2 litres of petrol. How far can it travel on 6 litres?
5. A tea-picker can fill 2 baskets in an hour. How many baskets can she fill in 8 hours?

11.7 continued

6. If 40 newspapers cost £18, how much will 10 newspapers cost?

7. 6 pieces of fence cost Rs720. Find the cost of 2 pieces.

8. A recipe for 8 people requires 1800g of aubergines. How many grammes of aubergines are required for 2 people?

9. A train travels 36 km in 21 minutes. How long does it take to travel 72 km?

10. Ten buckets can carry 90 litres of water. How many buckets are needed to fill a tank of 135 litres?

11. 5 coaches are need to transport 250 children to school. How many coaches are needed to transport 1250 children?

12. A map has a scale of 2cm for each kilometre. How may kilometres are represented by a distance of 9cm on the map?

13. A box of nuts holds 250g. How many boxes are needed for 2 kilograms? (1000g = 1kg)

14. A tea-picker can harvest 16kg of tea in two days. How many days will it take to harvest 128kg?

15. A farmer can sow three quarters of his field in 9 days. How long will it take him to sow the remaining part?

16. A swimmer completes 4 lengths of a 25 metre pool in 2 minutes. How long will it take to swim a kilometre at the same speed?

Chapter 12

The meaning of number

Ganesh said to his father, "You know the other week I asked you about how many numbers there are and one of your answers was that there are nine. You said this is the friendly answer and that with the nine numbers and zero we can make all the other numbers."

"Yes", replied his father, "I remember."

"My teacher at school said that we have ten numbers because we have ten fingers and that people first learnt to count on their fingers. Is that right?"

"That may be the case", said his father, "but there is another explanation which is very different."

"In the ancient teachings of the Veda there are hidden descriptions about the meaning of numbers and it has to do with the way the universe is made. You may have read that in ancient Greece people understood there to be four elements, earth, water, fire, air and a fifth, which we call space and which they called the receptacle. They thought of space as the receptacle because everything is located in space. But in the Veda there are clues that there are not just five elements but nine. The other four are much more subtle and are not usually recognised. Each of the nine numbers, from one to nine, stands for a basic element of creation. The creation is made up of these great elements. The nine and zero make ten and so this is the natural base for numbers."

"What are these great elements, father?" quizzed Ganesh.

His father sat down at his desk, opened a drawer and took out a folder. He pulled out a piece of paper with a carefully drawn diagram and handed it to Ganesh. The picture showed a circle with nine points. Ganesh had never seen it before.

"The Absolute or God, Brahman, stands at one," his father explained.

"Number one runs through every number, is every number and generates every number. The same is true with God and every creature. The Absolute runs through every creature, is every creature and creates every creature.

"The second element is called Hidden Nature. It is like a store-house where everything is kept ready waiting to be brought out and shown. You cannot see into the store-house but can know that it is there just like sitting in a dark room not being able to see

Chapter 12 The meaning of number

anything but knowing that it is full of objects. Think of wood ready to be burnt on a fire. Before it is put on the fire the flames are hidden within and then, when placed on the fire, out come the flames. Hidden Nature is like this.

"After that comes Manifest Nature. Listen to the birdsong in the trees outside. It is the nature of birds to sing and they will always do this. Everything has its own nature, what it is like. This element has three parts which are like energies, called Satva, Rajas and Tamas. Stava is like an energy of goodness, Rajas is the energy of activity and Tamas is the energy which brings things to rest.

"The fourth element is the Feeling of Existence. It is your sense of who you are, how you feel to be "I". Everyone feels themselves to be themselves. This is also the element of feelings, thoughts, ideas and emotions.

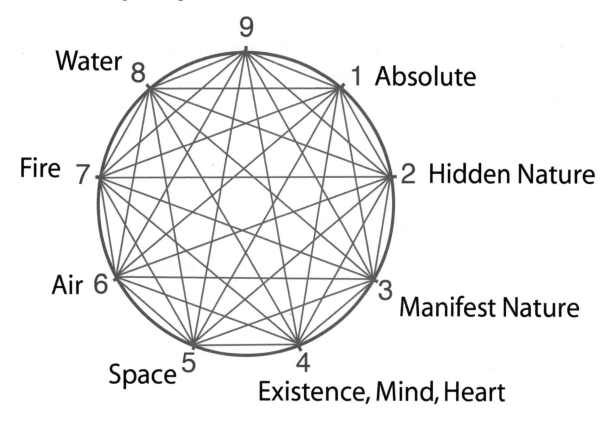

The Circle of Nine Points with the Great Elements of Creation

"At number five stands Space. You cannot see Space but you know it is there. It is the space in this room, the space outside this room and also the space inside the walls. It is everywhere and has no limit. It includes all the space in the universe with its stars and galaxies, planets and comets.

"The next great element is Air. It is the air we breath and it supports life. Air holds things up and supports life. It supports the aeroplanes in the sky, the birds when they fly and even makes the ice float in the cold seas. It is also the element of movement. Wherever there is movement, there is the presence of this element.

"Then comes Fire. This is the fire of the Sun and stars, light, colour, shape, warmth, electricity and energy. Everything on Earth contains this element because the Earth came from the sun. It also represent form or shape. All these things are part of this element.

"Water is at eight. This powerful element flows across the world. It gives rain and snow, the rivers and running streams, lakes and great oceans. Water feeds the plants and earth with the nutrients it carries. It is also the element of bonding which holds things together. Think of a leave from a tree. Once the water is taken out of the leave it turns to dust because the power of bonding is no longer there to hold the particles together.

The final element is Earth, which is the body of each creature. All the elements are contained in Earth. All the glittering crystals, held together with the bonds of Water are part of Earth. So too are all the chemical elements.

"Wow" said Ganesh. "That's an amazing description. I did not know about all this but it seems to make sense. What about zero?".

"Zero stands for the totality. It is mysterious yet very important. Think of an abacus with beads. When you count on an abacus you move the beads up one by one. The interesting step is when you have reached nine. After moving nine beads up in the units what is the next step? You move one up in the tens and then put all the nine units back to zero. The nine disappear into zero. So the zero stands for all nine when they are not showing themselves. When zero stands next to 1 it makes ten, 10. 1 and 10 are really the same unity - the difference is nothing!

"Number ten also stands for your real Self. Here is a story to illustrate.

"Ten people were going across the country to another land and they had to cross a river. The river was shallow but the currents were swift. They managed to cross the river, and after reaching the other shore, they wanted to make sure that no-one was drowned. Each of them lined the others up and found the total of nine only, for none of them would count himself. They were sorry and disturbed. A holy man was passing along the bank and seeing them miserable he asked the reason of their worries. They narrated their story. The holy man saw their difficulty and foolishness so he asked all of them to line up. With his stick he hit one and separated him from the others. The next one he hit twice, and then separated him. Likewise he hit the tenth man ten times and declared them ten and assured them that none was lost.

Chapter 12 The meaning of number

Ganesh liked the story very much but was still thinking about the elements.

"Father, I want to know more about these elements. Can you show me something of how they work?"

"Come outside." Replied his father. "Find something that takes your interest."

They walked out into the beautiful garden with its tall trees and deeply-scented flowers. Ganesh strolled around looking at the ground. He wandered over by a small group of trees and saw a dead mouse almost invisible under some fallen leaves. He gingerly picked it up by its tail and brought it over to where his father was standing.

"There!" he said. "I found this mouse. Can you tell me about the elements in this?"

His father took it from him and carefully laid it down on a large stone.

"It looks as if it's been dead for a few days. I am surprised the insects have not eaten more. You see it is decomposing. The body of this mouse is its Earth. This element is the substance of all matter. When the mouse died, its consciousness left. Its individual self was no longer there. Its Nature disappeared because it no longer behaved like a mouse. It stopped breathing and so the Air was gone. Its body grew cold because the Fire left it. The water that was in its blood left and it became dry. As the water leaves the body slowly turns to dust. In this way the elements leave and return to where they came from. So the air returns to the air around us; the water returns to the water around us, and so on. So when life leaves so the elements leave. When life comes the elements come. The great elements work like this, they come and they go.

"Usually people do not think of numbers in this way but they represent the elements. Take nine for example, which stands for the element Earth. Earth is the last element and there is nothing beyond. It is the perfect end to the creation. This perfection is reflected in the number 9. When you multiply 9 by any number the total of the digits is always 9. For example, $3 \times 9 = 27$ and $2 + 7 = 9$, $6 \times 9 = 63$ and $6 + 3 = 9$, $574 \times 9 = 5166$ and $5 + 1 + 6 + 6 = 18$, $1 + 8 = 9$. In this way you can never leave 9."

"What about zero, father?" he asked.

"Zero is special, very special" replied his father. "It is the nothingness into which everything disappears at the end. In mathematics it is highly significant and quite magical. Think of 5 rupees. Just by adding nothing you can change it into 500000 rupees! People did not trust it when first thought of but now everyone is used to it."

Ganesh wondered at all this but felt happy with the explanation.

Chapter 13

Algebra and Equations

Codes and substitution

For many centuries codes have been used to convey secret messages. All written codes are based on substituting the letters of a message for something else, which can be other letters, numbers or some other symbols.

Here is the key to a simple code based on substituting letters for two-digit numbers.

A	B	C	D	E	F	G	H	I	J	K	L	M
21	22	23	24	25	26	27	28	29	30	31	32	33

N	O	P	Q	R	S	T	U	V	W	X	Y	Z
34	35	36	37	38	39	40	41	42	43	44	45	46

Use this key to find the secret message below:

4028253825 2939 34354028293427 252940282538 27353524 343538

....

222124, 224140 4028293431293427 3321312539 2940 3935

....

Modern codes have many different uses and can be very sophisticated. For example, when shopping online bank details are encoded to ensure the safe transfer of information. Here is another simple code based on a grid.

Each letter is given a row and column reference. For example, B3 stands for H. Y and Z have the same box and can be got from context.

Chapter 13 Algebra and Equations

Using the grid on the previous page decode this message,

C3A1C1A5 A1 D4A5A3D3A5D5 A3C5A4A5 A1C4A4 B2B4E2A5 B4D5 D5C5 A1

....

B1D3B4A5C4A4 D5C5 D4A5A5 B4B1 D5B3A5E5 A3A1C4 A2D3A5A1C1 B4D5

....

Order of operations

To solve equations in algebra you need to have an understanding of the order of operations with numbers. An easy acronym is BIDMAS, which stands for Brackets, Indicies, Division, Multiplication, Addition and Subtraction. This tells you the correct order for combined operations.

For example, $4+2\times3=10$ or 18 according as to whether you multiply the 2 and 3 first and then add 4 or add 2 together and then multiply by 3. In BIDMAS, multiplication comes before addition and therefore the answer is 10.

Example $(6+2) \div (5-1) + 9$

$$(6+2) \div (5-1) + 9 = 8 \div 4 + 9 = 2 + 9 = 11$$

| 1. Deal with the brackets first. | 2. Division before addition. |

13.1 Use BIDMAS to find the correct order and answer these:t

1. $4+6\times2=$
2. $(4+6)\times2=$
3. $6+3\times4=$
4. $(6+3)\times4=$
5. $24\div6+2=$
6. $24\div(6+2)=$
7. $20\div4+6=$
8. $20\div(4+6)=$
9. $5+4\times3+3=$
10. $(5+4)\times(3+3)=$
11. $(5+4)\times3+3=$
12. $6+3\times4+2=$
13. $(6+3)\times4+2=$
14. $(6+3)\times(4+2)=$
15. $10+2\times5+5=$
16. $10+2\times(5+5)=$
17. $(10+2)\times5+5=$
18. $72\div(6+3)-1=$

The Curious Hats of Magical Maths

In algebra, the multiplication sign is not used much. For example, 3x means $3 \times x$. Numbers are placed before letters so you write 3x and not x3. This is part of algebraic syntax.

Substitution

Given values for the letters in an expression you can substitute those values into the expression to find its whole value.

Example Given that $x = 3$ and $y = 2$, find the value of $8x - 7y$

$$8x - 7y = 8 \times 3 - 7 \times 2 = 24 - 14 = 10$$

Example Given that $a = 2$ and $b = 4$, find the value of $2a^2b - 7b$

$$2a^2b - 7b = 2 \times 2^2 \times 4 - 7 \times 4 = 32 - 28 = 4$$

13.2 If $x = 5$, find the value of the following:

1. $4 + x =$
2. $x - 3 =$
3. $x - 1 + 8 =$
4. $3x + 2 =$
5. $4x \div 10 =$
6. $2x + x =$
7. $7x - 30 =$
8. $x + x + x =$
9. $x^2 - 1 =$
10. $2x \div 5 =$
11. $9x - 40 + x =$
12. $7x - 35 =$

13.3 If $a = 2, b = 3$ and $c = 4$, find the value of the following:

1. $a + b =$
2. $b + c =$
3. $a + c =$
4. $c - a =$
5. $b - a =$
6. $a + b + c =$
7. $a + b - c =$
8. $2a + b =$
9. $3b - c =$
10. $5c + a =$
11. $6c + 2a =$
12. $2c - 2a =$

13.4 Given that $x = 2, y = 1$, and $z = 0$, find the value of the following:

1. $2x + 3 =$
2. $3y - 2 =$
3. $11 - z =$
4. $2z + xy =$
5. $4x + 8z =$
6. $5z + 1 =$
7. $y \times y \times y =$
8. $x + y + z =$
9. $xz =$
10. $xyz =$
11. $6x - y + z =$
12. $y + 3x =$
13. $z + x - y =$
14. $8y - 4x =$
15. $9x - z =$
16. $3y + x - z =$
17. $7xy + 2x =$
18. $3x^2y =$
19. $x - 5yz =$
20. $5xy^2 - 9y =$

Chapter 13 Algebra and Equations

Simplifying by collecting like terms

In algebra, **numbers** are expressed as numbers or letters or mixtures of the two. For example, $2a$, 17 and x^2 are all numbers.

Terms are numbers separated by plus and minus signs or equal signs. For example, the expression, $x^2 + 3y - 4$ has three terms and the equation, $\frac{2x+1}{7} + 3 = 5x - 4$ has four terms.

Like terms are terms which are identical in every respect except for the number at the beginning. For example, $3ab$, $0.2ab$, $\frac{2}{5}ab$ and $-ab$ are all like terms.

Expressions can be simplified by collecting like terms.

Example Simplify $2x + 5x - x + 6x$

$$2x + 5x - x + 6x = 12x$$

Add the numbers, $2 + 5 - 1 + 6 = 12$

Example Simplify $3a + b - 4a + 7b + 2a$

$$3a + b - 4a + 7b + 2a = a + 8b$$

1. Scan across adding up the number of terms in a, $3 - 4 + 2 = 1$, giving 1a which is just written as a.
2. Add the number of terms in b, $1 + 7 = 8$ giving 8b. The answer is $a + 8b$.

13.5 Simplify by collecting like terms

1. $3p + 9p =$
2. $2a + 3a + 2a =$
3. $4b - 5b + 3b =$
4. $a + 2a + 3a =$
5. $8x + 3c + 5x =$
6. $11q - q =$
7. $a + b + 4a + b =$
8. $6x + y + x + y =$
9. $a + a + a =$
10. $3d + d - 4 =$
11. $2x + 4y + x + y =$
12. $n - 3 + n =$
13. $2x + 3 + x - 5 =$
14. $9 + 3x - 4x =$
15. $p \times p \times 7 =$

The Curious Hats of Magical Maths

At sight equations

Simple equations can be solved just by looking. The rule for this is *By Inspection*.

13.6 By Inspection

1. $x + 6 = 13$
   $x =$

2. $y - 4 = 5$
   $y =$

3. $8 + a = 14$
   $a =$

4. $15 = n + 7$
   $n =$

5. $3 = x - 3$
   $x =$

6. $b - 2 = 6$
   $b =$

7. $w + 4 = 7$
   $w =$

8. $0 = f - 3$
   $f =$

9. $19 - c = 13$
   $c =$

10. $12 = 21 - z$
    $z =$

11. $3x = 12$
    $x =$

12. $9b = 63$
    $b =$

13. $5a = 45$
    $a =$

14. $24 = 8y$
    $y =$

15. $16x = 32$
    $x =$

16. $3z = 9$
    $z =$

17. $2x + 1 = 5$
    $x =$

18. $3x + 1 = 19$
    $x =$

19. $2x + 3 = 13$
    $x =$

20. $5b - 1 = 9$
    $b =$

21. $p - 3 = 0$
    $p =$

22. $a + 3a = 40$
    $a =$

23. $9 = 8c - 5c$
    $c =$

24. $7x + 3 = 3$
    $x =$

25. $2m - 8 = 0$
    $m =$

26. $28 = 7 + 3b$
    $b =$

27. $3y - 18 = 0$
    $y =$

28. $7x - 30 = 5$
    $x =$

Chapter 13 Algebra and Equations

13.7 Solve these equation with fractions By Inspection

1. $\dfrac{x}{2} = 3$
   $x =$

2. $\dfrac{w}{5} = 2$
   $w =$

3. $3 = \dfrac{y}{4}$
   $y =$

4. $7 = \dfrac{n}{3}$
   $n =$

5. $\dfrac{100}{x} = 25$
   $x =$

6. $\dfrac{39}{y} = 3$
   $y =$

7. $\dfrac{32}{z} = 4$
   $z =$

8. $3 = \dfrac{9}{m}$
   $m =$

9. $\dfrac{6}{x} = 3$
   $x =$

10. $\dfrac{y}{3} = 4$
    $y =$

11. $3 = \dfrac{h}{20}$
    $h =$

12. $\dfrac{15}{g} = 3$
    $g =$

13. $6 = \dfrac{30}{a}$
    $a =$

14. $\dfrac{10}{b} = 2$
    $b =$

15. $\dfrac{m}{2} = 16$
    $m =$

16. $1 = \dfrac{n}{7}$
    $n =$

17. $\dfrac{35}{c} = 7$
    $c =$

18. $\dfrac{x}{13} = 0$
    $x =$

19. $5 = \dfrac{60}{y}$
    $y =$

20. $\dfrac{50}{w} = 2$
    $w =$

13.8 Mixed examples By Inspection

1. $6a = 24$
   $a =$

2. $m + 3 = 15$
   $m =$

3. $8 = 2c$
   $c =$

4. $5b = 40$
   $b =$

5. $n - 7 = 7$
   $n =$

6. $6 + p = 100$
   $p =$

7. $7 = d - 52$
   $d =$

8. $3z + 2 = 14$
   $z =$

9. $8 + h = 28$
   $h =$

10. $0 = y - 4$
    $y =$

11. $3x - 5 = 10$
    $x =$

12. $14 = 5 + b$
    $b =$

77

The Curious Hats of Magical Maths

Solving equations by Balance

An equation is like a balance. There is a sutra which simply says **When the total is the same it is zero**. This may seem a little cryptic but it reminds you that both sides of an equation are equal and that the difference is nought or zero. You can add, subtract, multiply or divide both sides of the equation by the same number and the balance is retained. You can think of this as *Do to one side as you do to the other*.

Example Solve $3x - 4 = 29$

1. The aim is to obtain x by itself. Add 4 to both sides to give $3x = 33$.

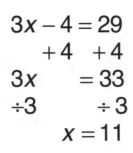

2. Divide both sides by 3. The solution is $x = 11$.

$$3x - 4 = 29$$
$$+4 \quad +4$$
$$3x \quad = 33$$
$$\div 3 \quad \div 3$$
$$x = 11$$

13.9 *Keep the equals signs in line*

1. $5x - 3 = 7$
2. $6x + 5 = 23$
3. $3 + 2x = 15$
4. $14 = 3x - 1$
5. $3x - 4 = 11$
6. $9x + 1 = 100$
7. $5 + 3x = 11$
8. $31 = 7x + 3$
9. $7x + 3 = 17$
10. $3x - 5 = 10$
11. $8 + 4x = 8$
12. $100 = 5x - 5$

Chapter 13 Algebra and Equations

Example $4x + 3 = 2x + 8$

$$
\begin{aligned}
4x + 3 &= 2x + 8 \\
-2x \quad &\quad -2x \\
2x + 3 &= 8 \\
-3 \quad &\; -3 \\
2x \quad &= 5 \\
\div 2 \quad &\; \div 2 \\
x &= 2\tfrac{1}{2}
\end{aligned}
$$

1. x terms appear twice. Subtract the smaller term from both sides to give $2x + 3 = 8$.

2. Subtract 3 from both sides.

3. Divide both sides by 2.

13.10 Keep the equals signs in line

1. $5x + 1 = 3x + 7$
2. $7x + 4 = 2x + 19$
3. $a + 2 = 3a - 4$
4. $3b - 4 = 2b + 1$
5. $8y - 3 = 2y + 3$
6. $9x = 4x + 15$
7. $11m = 5m + 36$
8. $28 + 2p = 9p$
9. $5d + 12 = 17d$
10. $3c = 7c - 12$
11. $5z - 2 = 2z + 10$
12. $8x - 9 = 4x - 1$
13. $6a - 11 = a - 1$
14. $7c + 1 = 5c + 1$
15. $3b + 5 = 20b - 29$

The Curious Hats of Magical Maths

Equations with fractions

The easiest way to solve equations with fractions is usually to multiply both sides of the equation by the denominator. This will then leave a simple equation which can be solved as before.

Example $\dfrac{2x}{3} = 20 - x$

$$\dfrac{2x}{3} = 20 - x$$
$$\times 3 \qquad \times 3$$
$$2x = 60 - 3x$$
$$+3x \qquad +3x$$
$$5x = 60$$
$$x = 12$$

1. 3 is the denominator so multiply both sides by 3. This results in a simple equation, $2x = 60 - 3x$.

2. Add $3x$ to both sides.

3. Divide both sides by 5.

Sometimes it is best to add or subtract a term on both sides first.

Example $4 + \dfrac{2x}{5} = 8$

$$4 + \dfrac{2x}{5} = 8$$
$$-4 \qquad -4$$
$$\dfrac{2x}{5} = 4$$
$$\times 5 \quad \times 5$$
$$2x = 20$$
$$x = 10$$

1. Subtract 4 from both sides.

2. Multiply both sides by 5.

3. Observation gives $x = 10$.

13.11 Keep the equals signs in line

1. $\dfrac{3x}{2} = 12$

2. $\dfrac{2x}{3} + 1 = 5$

3. $2 + \dfrac{x}{3} = 7$

4. $\dfrac{4x}{3} - 5 = 11$

5. $\dfrac{x}{7} + 2 = 5$

6. $3 = 1 + \dfrac{x}{2}$

7. $7 = \dfrac{2x}{9} - 5$

8. $\dfrac{8x}{3} - 8 = 2x$

13.12 Equations with the unknown on both sides

1. $5n = 3n + 10$
2. $3a = a + 2$
3. $4b = b + 15$
4. $7c = c + 12$
5. $4m = m + 30$
6. $12p = p + 66$
7. $10h = 3h + 21$
8. $7n + 1 = 6n + 8$
9. $4d + 3 = d + 9$
10. $3v + 7 = v + 15$
11. $6w - 1 = 3w + 8$
12. $5a - 4 = 2a + 5$
13. $1 + 3e = e + 7$
14. $4y - 11 = 2y + 11$
15. $1 + 5c = 3c + 13$
16. $6p = 3p + 24$
17. $13g = 7g + 24$
18. $5t = 8 + t$
19. $2n = 7 + n$
20. $3x - 3 = 2x - 1$
21. $12a - 3 = a + 19$

Chapter 14

A few short cuts for multiplying

Squaring numbers ending in 5

This very easy method for squaring numbers ending in 5 uses the *By one more than the one before* rule.

Example 65^2

$$65^2 = 42/25$$

1. The one before the 5 is 6 and one more than this is 7. Multiply the 6 and 7 together to give 42.

2. Put down the square of 5 which is 25. The answer is 4225.

14.1

1. $35^2 =$
2. $55^2 =$
3. $25^2 =$
4. $45^2 =$
5. $15^2 =$
6. $95^2 =$
7. $85^2 =$
8. $75^2 =$
9. $105^2 =$

How does it work?

The diagram below shows a square of side 35. The area of this square is $35^2 = 1225$. By moving the top rectangle round to the side you can see why 30 and 40 are multiplied together.

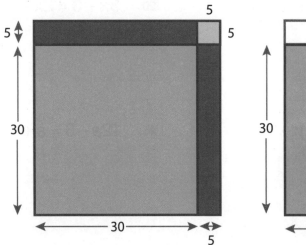
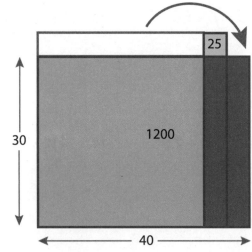

Chapter 14 A few short cuts for multiplying

You can extend this to the same numbers but with zeros on the end. The method is exactly the same and you just need to put double the number of zeros on the end.

Example 7500^2

$$7500^2 = 56250000$$

1. $7 \times 8 = 56$ and $5^2 = 25$.

2. 7500 has two zeros and so the answer has four.

14.2

1. $450^2 =$
2. $650^2 =$
3. $350^2 =$
4. $550^2 =$
5. $150^2 =$
6. $750^2 =$
7. $9500^2 =$
8. $4500^2 =$
9. $8500^2 =$

When the final digits add to 10

This is a sub-sutra or sub-rule of *By one more than the one before* and is almost the same as for squaring numbers ending in 5. This can be used for multiplying numbers whose last digits add to 10 and whose first digits are the same.

Example 47×43

$$47 \times 43 = 20 / 21$$

1. 4 is the first digit and one more than this is 5. $4 \times 5 = 20$.

2. $7 \times 3 = 21$. The answer is 2021.

14.3 Use this method to work these out

1. $36 \times 34 =$
2. $62 \times 68 =$
3. $27 \times 23 =$
4. $18 \times 12 =$
5. $17 \times 13 =$
6. $16 \times 14 =$
7. $24 \times 26 =$
8. $98 \times 92 =$
9. $74 \times 76 =$

The Curious Hats of Magical Maths

14.4 *Here are a few more*

1. $63 \times 67 =$
2. $57 \times 53 =$
3. $38 \times 32 =$
4. $46 \times 44 =$
5. $36 \times 34 =$
6. $43 \times 47 =$
7. $54 \times 56 =$
8. $82 \times 88 =$
9. $102 \times 108 =$

Multiplying by 11

This is also very easy and people are often quite surprised by it. It uses the *Osculation* rule! *Osculate* mean to kiss! The idea is that the numbers standing next to each other kiss, or touch, and produce their sum.

Example 43×11

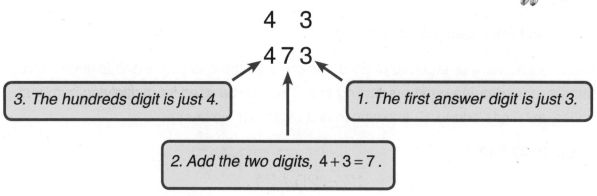

3. The hundreds digit is just 4.

2. Add the two digits, $4 + 3 = 7$.

1. The first answer digit is just 3.

Example 79×11

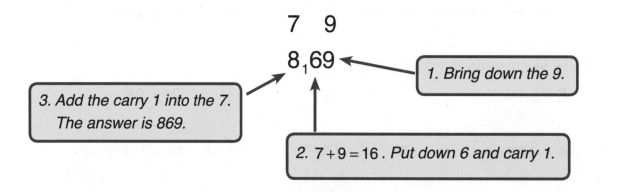

3. Add the carry 1 into the 7. The answer is 869.

1. Bring down the 9.

2. $7 + 9 = 16$. Put down 6 and carry 1.

84

14.5 Multiply by 11

1. $23 \times 11 =$
2. $42 \times 11 =$
3. $33 \times 11 =$
4. $51 \times 11 =$
5. $15 \times 11 =$
6. $26 \times 11 =$
7. $81 \times 11 =$
8. $63 \times 11 =$
9. $11 \times 11 =$
10. $50 \times 11 =$
11. $67 \times 11 =$
12. $38 \times 11 =$

This can easily be extended to multiplying numbers of any length by 11.

Example 2563×11

$$2563$$
$$28{,}193$$

1. 3 is brought down. $3+6=9$.
2. $6+5=11$. Put down 1 and carry 1.
3. $5+2+1=8$. Bring down 2 for the final digit.

14.6 Multiply by 11

1. $421 \times 11 =$
2. $133 \times 11 =$
3. $234 \times 11 =$
4. $701 \times 11 =$
5. $513 \times 11 =$
6. $400 \times 11 =$
7. $520 \times 11 =$
8. $333 \times 11 =$
9. $273 \times 11 =$
10. $785 \times 11 =$
11. $1321 \times 11 =$
12. $2104 \times 11 =$
13. $6574 \times 11 =$
14. $5073 \times 11 =$
15. $8294 \times 11 =$

Chapter 15

Digital Roots

Adding the digits of a number

When you add the digits of a number and keep adding until there is only one number left that number is called the digital root. Some examples are set out below.

Number	Adding	Digital Root
14	1+4=5	5
48	4+8=12, 1+2=3	3
241	2+4+1=7	7
5643	5+6+4+3=18, 1+8=9	9

In the second example, 48, 4 add 8 makes 12. Because 12 has two digits add the 1 and 2 to make 3. This is the digital root.

The rule for this is, *By Addition and Subtraction* and this is because as you add the digits together you subtract the number of digits. Here is the hat.

15.1 *Find the digital root (DR)*

1. 23 DR =
2. 26 DR =
3. 12 DR =
4. 35 DR =

5. 39 DR =
6. 95 DR =
7. 87 DR =
8. 68 DR =

9. 123 DR =
10. 245 DR =
11. 635 DR =
12. 409 DR =

Chapter 15 Digital Roots

15.2 *Find the digital root*

1. 61   DR =
2. 72   DR =
3. 44   DR =
4. 24   DR =
5. 11   DR =
6. 77   DR =
7. 86   DR =
8. 93   DR =
9. 64   DR =
10. 672   DR =
11. 594   DR =
12. 911   DR =
13. 216102   DR =
14. 912432   DR =
15. 999999   DR =

The digital root of a number is also the remainder when that number is divided by nine. Examples:

$$12 \div 9 = 1 \text{ rem } 3, \qquad 32 \div 9 = 3 \text{ rem } 5,$$
$$1 + 2 = 3 \qquad\qquad\qquad 3 + 2 = 5$$

You will have come across this previously when dividing by 9 in the chapter on Nikhilam Division.

What happens when the remainder is 0? For example, $27 \div 9$ is 3 remainder 0, but the digital root of 27 is 9. All you need to do is remember that a remainder of 0 shows up as a digital root of 9.

15.3 *Divide each number by 9 and put a circle around the remainder*

1. $14 \div 9 =$
2. $17 \div 9 =$
3. $13 \div 9 =$
4. $10 \div 9 =$
5. $43 \div 9 =$
6. $34 \div 9 =$
7. $32 \div 9 =$
8. $68 \div 9 =$
9. $123 \div 9 =$
10. $245 \div 9 =$
11. $231 \div 9 =$
12. $345 \div 9 =$

The Curious Hats of Magical Maths

Circle of 9 Points

The circle of 9 points can be used to show the patterns of digital roots. Here is the circle with every point joined to every other point.

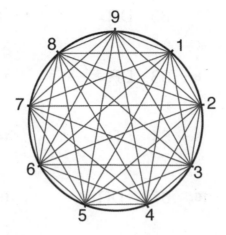

Copy the pattern on using a pencil and ruler to join each point to every other point. Colour the pattern in.

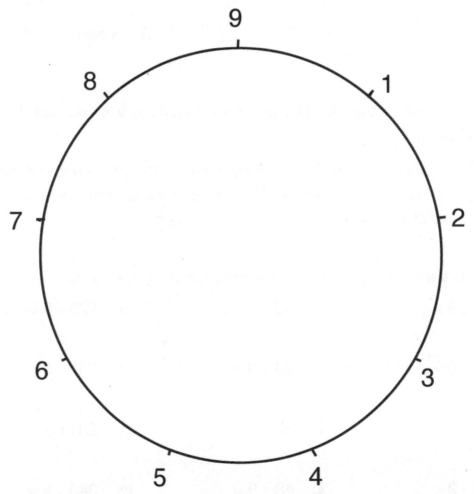

Chapter 15 Digital Roots

Digital root patterns

15.4 *Complete the four times table below. Find the digital root of each answer and enter them in the table.*

4 X table		Digital Root		4 X table	4 X table	Digital Root
1 × 4 = 4						
2 × 4 = 8						
3 × 4 = 12	1+2 = 3					
				12 × 4 = 48		

The pattern of digital roots is, 4, 8, 3, 7, 2, 6, 1, 5, 9. After that it repeats. If you continue the times table the pattern will carry on repeating.

Plotting the pattern on the circle of nine points

To see the pattern in these numbers you can plot them on the circle of nine points. To do this, start at 4 and draw a straight line from 4 to 8. Then draw a straight line from 8 to 3, the next number. In the same way draw lines from 3 to 7, 7 to 2, 2 to 6, 1 to 5, 5 to 9 and 9 to 4. The resulting pattern is shown below.

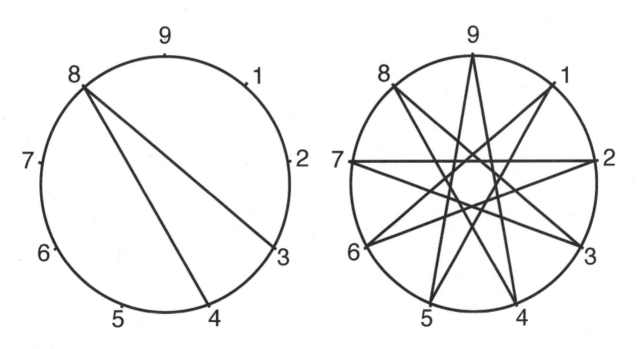

The Curious Hats of Magical Maths

15.5 *Complete the chart for the 2 X table and draw the pattern on the circle*

2 X table	Digital Root

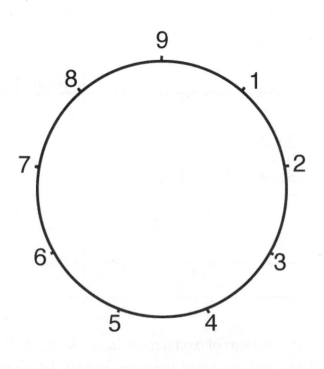

15.6 *Now do the same for the 3 X table and draw the pattern on the circle*

3 X table	Digital Root

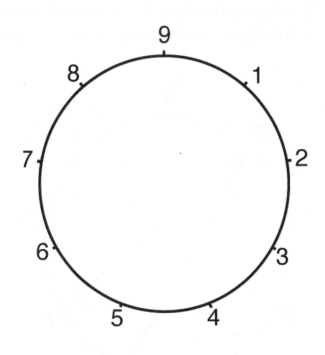

15.7 *Now do the same for the 5 X table*

5 X table	Digital Root

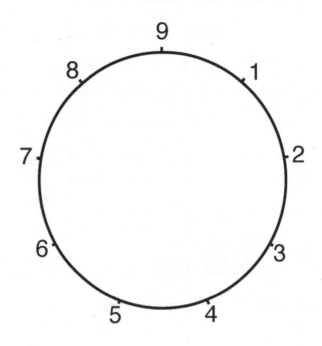

15.8 *This is for the 6 X table*

6 X table	Digital Root

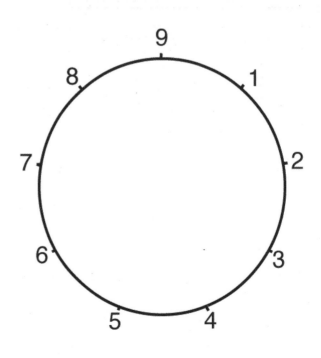

The Curious Hats of Magical Maths

15.9 *7 X table*

7 X table	Digital Root

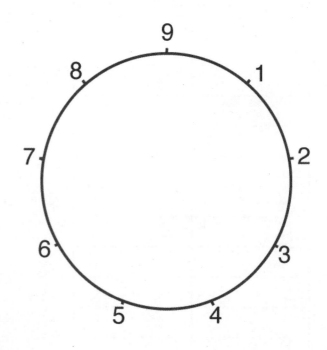

15.10 *8 X table*

8 X table	Digital Root

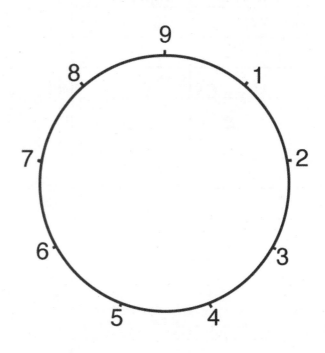

15.11 Each times tables has the same pattern as one other. Complete the following statements:

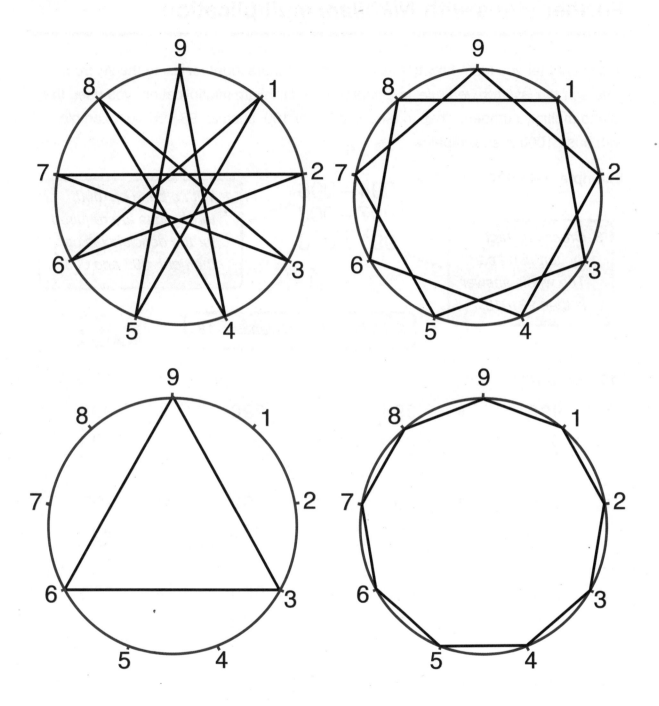

Chapter 16

Further steps with *Nikhilam* multiplication

In this chapter you learn how to multiply large numbers together using the *All from nine and the last from ten rule*. In Chapter 4, on Nikhilam multiplication, you used this rule to multiply numbers close to a base of 10 or 100, such as 98 96. An example with base 1000 is shown below.

Example 994 × 997

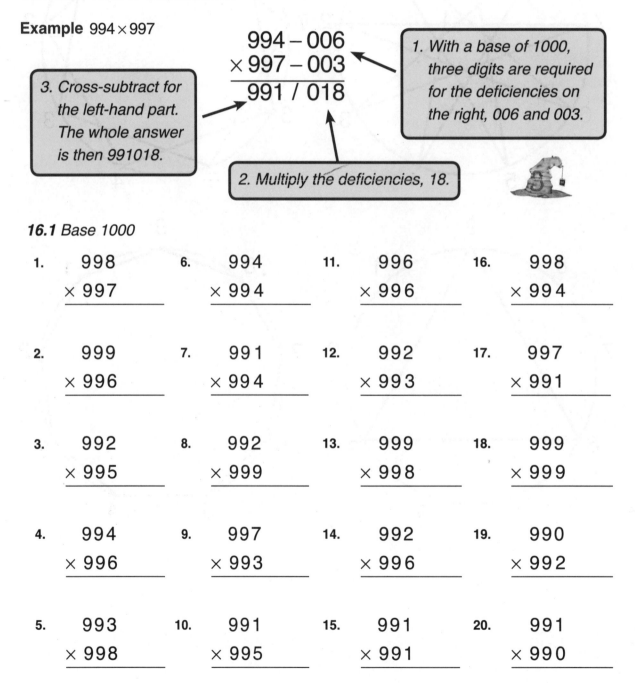

1. With a base of 1000, three digits are required for the deficiencies on the right, 006 and 003.

2. Multiply the deficiencies, 18.

3. Cross-subtract for the left-hand part. The whole answer is then 991018.

16.1 Base 1000

1. 998 × 997
2. 999 × 996
3. 992 × 995
4. 994 × 996
5. 993 × 998

6. 994 × 994
7. 991 × 994
8. 992 × 999
9. 997 × 993
10. 991 × 995

11. 996 × 996
12. 992 × 993
13. 999 × 998
14. 992 × 996
15. 991 × 991

16. 998 × 994
17. 997 × 991
18. 999 × 999
19. 990 × 992
20. 991 × 990

94

Chapter 16 Further steps with Nikhilam multiplication

As long as one of the numbers is close to the base this method is very easy.

16.2 *This exercise includes carrying to the left.*

1.	887 × 999	9.	667 × 997	17.	405 × 993		
2.	936 × 993	10.	487 × 998	18.	628 × 997		
3.	788 × 997	11.	599 × 996	19.	774 × 994		
4.	732 × 998	12.	482 × 993	20.	308 × 998		
5.	659 × 999	13.	259 × 997	21.	709 × 994		
6.	932 × 992	14.	386 × 991	22.	354 × 995		
7.	600 × 998	15.	247 × 996	23.	291 × 990		
8.	846 × 996	16.	356 × 992	24.	456 × 995		

Above the base

Example 1007 × 1003

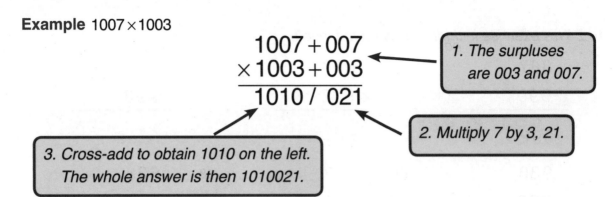

1. The surpluses are 003 and 007.
2. Multiply 7 by 3, 21.
3. Cross-add to obtain 1010 on the left. The whole answer is then 1010021.

16.3

1. 1003 × 1002
2. 1004 × 1006
3. 1007 × 1008
4. 1008 × 1004
5. 1007 × 1006
6. 1007 × 1007
7. 1009 × 1006
8. 1011 × 1006
9. 1008 × 1003
10. 1008 × 1008
11. 1003 × 1005
12. 1001 × 1007
13. 1004 × 1004
14. 1009 × 1005
15. 1001 × 1001
16. 1722 × 1002
17. 1834 × 1003
18. 1920 × 1005
19. 1442 × 1007
20. 1253 × 1009
21. 1300 × 1004

Chapter 16 Further steps with Nikhilam multiplication

Here's a short rhyme to help you remember about the digits in the right-hand part,

*The complement digits in every case
equals the number of noughts in the base.*

16.4 Mixed practice

1. 992 × 993

2. 997 × 995

3. 1009 × 1005

4. 999 × 999

5. 996 × 991

6. 1003 × 1008

7. 992 × 997

8. 994 × 990

9. 1004 × 1009

10. 938 × 997

11. 1010 × 1007

12. 945 × 995

13. 968 × 998

14. 1021 × 1004

15. 1032 × 1003

16. 497 × 996

17. 1023 × 1002

18. 1001 × 1076

19. 500 × 994

20. 1123 × 1003

21. 1235 × 1009

Chapter 17

Working with decimal fractions

There are three distinct features about our number system. The first is that it is *decimal* which means that it has a base of 10. The second is that *zero* is used if there is no number. The third is *place value* which means that a number like 37 is different from 73 because of the position of the digits.

Decimal fractions are used for numbers as tenths, hundredths, thousandths and so on and to distinguish a whole number from a decimal fraction a decimal point is used. For example, 43.17 is 4 tens, 3 units, 1 tenth and 7 thousandths. So numbers to the right of the decimal point stand for parts.

Sometimes, in life, it is easy to forget the parts of things and think that we deal with the whole when, in fact, we do not. Here is a short story to illustrate this.

There was an impatient man in India who wanted to realise God. So he went to a holy man to ask for help. The holy man said that all he needed to do was to remember that God is in everything and means no harm and then he would realise God. So the man happily went on his way busily remembering that God is in everything.

On his way home he was walking down a narrow country lane with high hedges on either side. He suddenly saw an elephant coming in the opposite direction. Sitting on top of the elephant was a mahout, an elephant driver, who, seeing the man cried out, "Get out of the way!"

Seeing that there was no room to get past the man said to himself, "God is in me, God is in the elephant. Can God harm God? No!". And he carried on walking towards the elephant. The elephant driver again called for him to get out of the way and again the man said to himself, "God is in me, God is in the elephant. Can God harm God? No!", and carried on going. As soon as the elephant reached the man it picked him up in his trunk and threw him over the hedge and he landed in a muddy field.

Hurt and sorrowful the man returned to the holy teacher and told him of his experience. The holy man said, "You were quite right to remember that God is in you and God is in the elephant, but God is also in the elephant driver and he told you to get out of the way!".

This story illustrates how easy it is to forget the whole by separating things off into parts.

Chapter 17 Working with decimal fractions

Most basic calculations with decimal fractions are performed in the same way as for ordinary numbers. For both addition and subtraction keep the decimal points in a vertical line.

17.1 Decimal additions

1. 2.4
   + 3.2

2. 12.3
   + 7.8

3. 84.5
   + 27.9

4. 8.2
   + 19.9

5. 4.36
   + 2.74

6. 0.23
   + 7.29

7. 0.07
   + 0.56

8. 0.364
   + 0.88

9. 45.32
   + 78.47

10. 82.34
    + 90.39

11. 13.07
    + 38.08

12. 77.32
    + 14.86

13. 846.3
    + 64.8

14. 34.7
    + 640.8

15. 984.5
    + 675.7

16. 85.4
    + 714.8

17.2 Use either On the flag or Nikhilam for these subtractions

1. 7.3
   − 4.8

2. 72.3
   − 48.5

3. 24.6
   − 12.8

4. 34.5
   − 16.9

5. 50.4
   − 8.7

6. 37.5
   − 13.6

7. 48.6
   − 39.2

8. 71.2
   − 40.7

9. 2.42
   − 0.77

10. 0.78
    − 0.29

11. 354.3
    − 187.9

12. 323.4
    − 186.6

13. 736.3
    − 45.7

14. 5.45
    − 2.78

15. 9.43
    − 1.78

16. 6.432
    − 3.786

The Curious Hats of Magical Maths

Nikhilam multiplication with decimals

Example 8.9 × 0.97

```
  89 – 11
× 97 – 03
─────────
  8.6   33
```

1. Treat this as 89 × 97. The deficiencies are 11 and 3. Multiplying on the right gives 33 and cross-subtraction gives 86.

2. Count the decimal digits - three. This tells you where to place the point in the answer; there must be three digits following the point.

17.3 *Set these out as shown above*

1. 9.6 × 9.4
2. 8.8 × 9.8
3. 0.95 × 9.2
4. 93 × 0.91
5. 0.94 × 0.97
6. 8.7 × 0.98
7. 1.2 × 1.3
8. 1.4 × 0.13
9. 10.2 × 10.4
10. 11.3 × 1.03
11. 0.104 × 1.12
12. 107 × 1.03
13. 96.3 × 9.98
14. 0.998 × 98.3
15. 98.9 × 99.5
16. 9.67 × 0.999
17. 0.987 × 0.994
18. 0.123 × 1.07

Chapter 17 Working with decimal fractions

Vertically and Crosswise multiplication with decimals

You can use the Vertically and Crosswise method for multiplying any two decimal numbers together. The knack is to multiply the numbers together as if there are no decimal points and then insert the point in the correct position in the answer. This is done most easily by counting the decimal digits in the two numbers. It is equal to the number of decimal digits in the answer.

Example 2.3×0.031

$$\begin{array}{r} 2\,3 \\ \times\ 3\,1 \\ \hline 0.07_1 3 \end{array}$$

1. Set it out as for 23×31.

2. Multiply by Vertically and Crosswise to give 713.

3. The total decimal digits in 2.3 and 0.031 is four. There must therefore be four digits after the point in the answer, 0.0713.

17.4 *Set these out as shown above*

1. 3.2×2.3
2. 1.4×65
3. 4.5×8.2
4. 0.12×7.3

5. 37×0.48
6. 0.81×0.24
7. 0.077×2.8
8. 3.1×0.71

9. 0.83×0.25
10. 92×0.22
11. 6.7×0.30
12. 6.7×0.03

The Curious Hats of Magical Maths

17.5 Set each problem out in the box on the right

1. Find the cost of 2.5 kg of butter at Rs72 per kilogram.

2. Curtain material costs Rs630 per metre. How much does 8.2m cost?

3. A car petrol tank holds 54 litres. If the car averages 9.2 km per litre, how many kilometres can it go on one full tank?

4. Sections of fence are 2.2m long. How many metres in length would sixteen sections cover?

5. A man earns £9.50 per hour. How much does he earn in a day assuming he works for 8.5 hours?

6. Find the area of sheet of paper measuring 28cm by 19cm.

7. Find the cost of 54 litres of petrol at Rs83 per litre.

8. A rectangular floor is 6.5m long and 4.2m wide. What is its area?

Chapter 17 Working with decimal fractions

Decimal division with single digit divisors

When dividing a decimal number by a single units digit the point stays in line. If the divisor itself is a decimal then use Proportionately to change both numbers so that the divisor is a units digit.

Example $3.428 \div 0.4$

$$3.428 \div 0.4 = 34.28 \div 4$$

1. 0.4 needs to be multiplied by 10 to make it into the units digit, 4. Do the same to 3.428. The simplest process is to move the decimal point one place to the right in both the numbers.

$$4 \overline{\smash{)}34._2 2_2 8}$$
$$8.\,5\,7$$

2. You can now divide easily by keeping the decimal points in line.

17.6 Use Proportionately to change the divisors into units digits

1. $0.64 \div 0.4$
2. $1.46 \div 0.2$
3. $48.3 \div 0.7$
4. $28.38 \div 0.2$
5. $6.065 \div 0.5$
6. $9.12 \div 0.8$
7. $3.059 \div 0.07$
8. $1.4 \div 0.8$
9. $6.204 \div 0.03$
10. $5.67 \div 0.02$
11. $6.782 \div 0.004$
12. $13.206 \div 0.06$

The Curious Hats of Magical Maths

Chapter 18

Puzzles and problems

This chapter consists of puzzles and problems under the headings of various sutras. The are often several ways of getting to the answer of a problem or puzzle and the headings are there simply to suggest that one method involves using that particular sutra. You can solve them in any way you like.

Proportionately

1. In music, a demisemiquaver is half of half of half a crotchet, and there are four crotchets in a semibreve. How many demisemiquavers are there in a semibreve?

2. The White Rabbit has an appointment to see the Red Queen at 4pm every day apart from weekends. On Monday, he arrives 32 minutes late. Each day after he hurries more and more and so manages to halve his lateness each day. On what day of the week does he arrive just 15 seconds late?

3. At the Bangalore Big Bake-off, prize money is awarded for 1st, 2nd and 3rd places in the ratio 3 : 2 : 1. Last year Mrs Dave and Mrs Chohan shared third prize equally. What fraction of the total prize money did Mrs Dave receive?

4. A square is divided into three identical rectangles. The middle rectangle is removed and placed on the side as shown. What is the ratio of the length of the perimeter of the square to the length of the perimeter of the resulting shape?

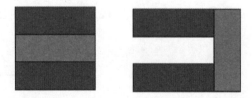

Chapter 18 Puzzles and Problems

5. All old Mother Hubbard had in her cupboard was a Giant Elephant chocolate bar. She gave each of her children one-twelfth of the chocolate bar. One third of the bar was left. How many children did she have?

....

6. Peter Piper picked a peck of pickled peppers. 1 peck = $\frac{1}{4}$ bushel and 1 bushel = $\frac{1}{9}$ barrel. How many more pecks must Peter Piper pick to fill a barrel?

....

7. Which of these fractions is closest to 1? $\frac{12}{23}$ $\frac{23}{34}$ $\frac{34}{45}$ $\frac{45}{56}$ $\frac{56}{67}$

....

8. 48% of the pupils at a certain school are girls. 25% of the girls and 50% of the boys at this school travel to school by bus. What percentage of the whole school travel by bus?

....

9. If an exam paper weighs 6 grams what is the weight of 140,000 exam papers?

....

10. A bottle contains 750 ml of mineral water. Ravi drinks 50% more than Harinder, and these two friends finish the bottle between them. How much does Ravi drink?

....

11. A ball is dropped onto a hard surface. Each time it bounces, it rebounds to exactly one third of the height from which it fell. After the second bounce the ball rises to a height of 12 cm. From what height was it originally dropped?

....

The Curious Hats of Magical Maths

12. A cyclist sets out on a long uphill ride. At 1pm he is one third of the way up the hill and at 3pm he is three quarters of the way up. What time did he set out?

13. Five identical rectangles fit together as shown. What is their total area?

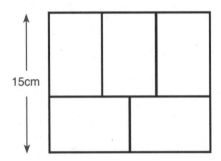

When it is the same, it is nought

This sutra carries the sense of when both sides of something are the same then the difference is nothing. A simple application is in symmetry. Reflective symmetry is when a mirror line can be drawn through a shape so that the two sides are identical. This can commonly be found in nature, for example with butterfly wings.

1. How many letters in the word MATHEMATICS have lines of symmetry?

2. Shade in boxes so that the shapes each have two lines of symmetry at right angles to each other.

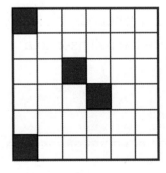 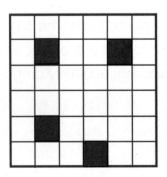 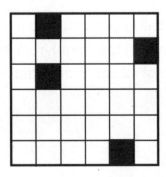

Chapter 18 Puzzles and Problems

3. These two pictures have 19 small differences. Mark the differences on the left-hand picture.

4. How many of the six faces of a die (shown below) have fewer than three lines of symmetry?

....

5. The diagram shows two arrows drawn on separate 6cm by 6cm grids. One arrow points North and other points West. When the two arrows are drawn on the same grid, still pointing in the same direction they overlap. What is the area of overlap?

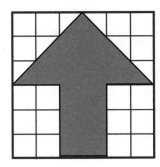 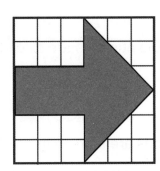

....

6. I am three times older than my daughter. In ten years time I will be twice as old as her. How old are both of us now?

....

The Curious Hats of Magical Maths

By Elimination and Retention

This sutra has not been used in this book yet but does come up in Book 2. The idea is that you eliminate something and keep something to help solve the problem. In the first two questions below, involving the counting up of shapes, you will find that you give your attention to one shape and count, momentarily eliminating the others, and then move on to the next shape. This is a good indication of how the sutra works.

1. How many triangles (of all types) are there in this shape?

....

2. How many rectangles are there?

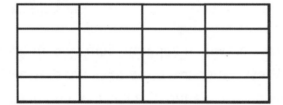

....

3. Three positive whole numbers are all different and their sum is 7. What is their product?

....

4. How many hexagons are there in the diagram?

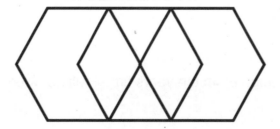

....

5. Patrick O'Connell, the farmer has three fields. In the largest he keeps 64 sheep. In the second largest he grows wheat and in the smallest he grows oranges. The areas of the three fields are in the ratio 2 : 3 : 4. Each year he harvests about 240 tons of wheat and 125 tons of oranges. Each year in April he sells the wool from the sheep at the wool market. The field with wheat has an area of 9000 m^2. What are the areas of the other two fields?

6. Each of the nine squares in the diagram are to be filled so that each row and each column contain one 1, one 2 and one 3 in some order. What must P + Q be?

Transpose and Adjust

This sutra has not been used previously in this book but comes up time and again later on. It has all sorts of uses, such as in transposing terms in equations and applying patterns. Whenever you mentally move one thing to another place in order to answer the problem is also an application of *Transpose and adjust*. Solving the following problems will give you a sense of its meaning.

1. What fraction of these shapes is shaded?

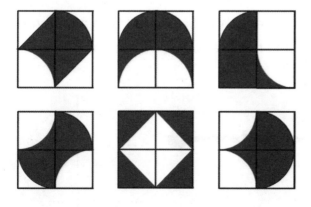

109

The Curious Hats of Magical Maths

2. A man has two sons. The sons are twins; they are the same height. If you add the man's height to the height of one son, you get 10 feet. The total height of the man's height and the two sons is fourteen feet.

What are the heights of the man and his sons?

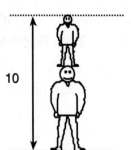
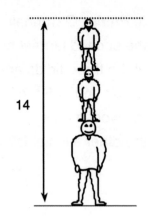

3. Each square has side length 1 unit. What is the area of triangle *PQR*, in square units?

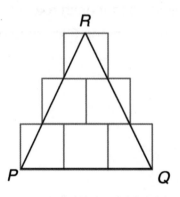

4. Use the diagram to find the formula for the area of a parallelogram.

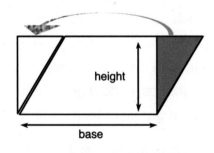

5. A rod has five equally spaced points numbered 1 to 5. The rod is rotated three times through 180°, first about 1, then about 2 and finally about 5. Whish point finishes in the same position as it was at the start?

Chapter 18 Puzzles and Problems

6. What area of these shapes is shaded?

....

7. What fraction of this shape is shaded? If the pattern were to continue for ever, what fraction would be shaded?

....

111

The Curious Hats of Magical Maths

8. What fraction of this shape is shaded yellow? Hint: You may find it useful to make a copy and cut out the pieces.

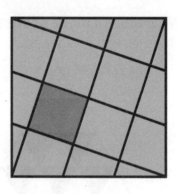

9. There was a problem weighing the baby at the clinic. The baby would not keep still and made the weighing scales wobble. So I held the baby and stood on the scales while the nurse read off 78 kg. Then the nurse held the baby while I read off 69 kg. Finally, I held the nurse while the baby read off 137 kg. What was the weight of the baby?

10. This design is formed by drawing six lines inside regular hexagon. The lines divide each side into three equal parts. What fraction of the whole shape is shaded?

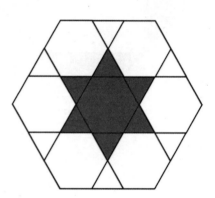

11. The parallelogram *ABCD* on the right has been divided into nine smaller parallelograms. The perimeters, in centimetres, of four of the smaller parallelograms are shown. The perimeter of *ABCD* is 21 cm. What is the perimeter of the green parallelogram?

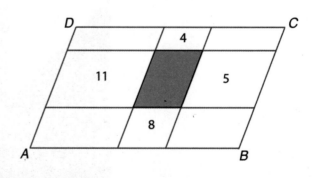

Appendix

The Sutras of Vedic Mathematics

The sutras, or rules, you have learnt about in this book are originally in the ancient Indian language of Sanskrit. The oldest scriptures and hymns in the world are written in this language, together with the great epic stories of the Ramayana and Mahabharata which tell the adventures of Rama and Arjuna and how they came to liberation. There are four original Veda as well as a number of supporting texts.

In the first half of the 20th Century a spiritual teacher named Shankaracarya Bharati Krishna Tirthaji studied these ancient Vedic texts and from them reconstructed the Vedic Maths sutras. There are sixteen main sutras and a similar number of sub-sutras. He wrote sixteen volumes about them but, unfortunately, these were lost before publication. Late in his life he wrote a single book illustrating the use of the sutras for solving a variety of problems. His work, Vedic Mathematics, was published in 1965. In his description of Veda he pointed out that there are two meanings. One is the collection of texts, such as the Rigveda, and this is the meaning commonly accepted. The second meaning is real knowledge which is found in the present moment. It is this meaning that is most appropriate to learning Vedic mathematics. The sutras are like seeds of knowledge that can grow in many different ways.

Bharati Krishna Tirthaji's aim was to enable people to be as happy as possible. In mathematics this meant finding ways in which problems could be solved with the minimum amount of effort, usually by mental working, and in a way that is both magical and delightful.

This book introduces ten of the sutras and a few of their applications. Each sutra has many different uses and they are very flexible.

A sutra is a short statement giving a principle, rule, pattern of working or aide memoir. The word *sutra* is a Sanskrit word meaning thread of knowledge and so each rule contains the knowledge required to solve particular problems. The sutras are really like the different blades on a Swiss Army penknife. Just as you can pull out whichever blade you need to do a job, so too, you can use whichever sutra is most suited to the task in hand.

The Curious Hats of Magical Maths

Here is a list of the sutras used in this book.

एकाधिकेन पूर्वेण Ekādhikena Purveṇa	By one more than the one before pp 18-24, 82-83	
यावदूनम् Yāvadūnam	Deficiency pp 14-17	
निखिलं नवतश्चरमं दशतः Nikhilam Navataścaraman Daśatah	All from 9 and the last from 10 pp 1-7, 9-13, 25-29, 46-52, 53-62, 94-97, 99-100	
ऊर्ध्व तिर्यग्भ्याम् Ūrdhva Tiryagbhyām	Vertically and crosswise pp 35-41, 101-102	
ध्वजाङ्क Dhvajānka	On the flag pp 42-45	
आनुरूप्येण Ānurupyena	Proportionately pp 63-67, 104-109	
वेष्टनम् Veshtanam	Osculation pp 84-85	
संकलन व्यावकलनाभ्याम् Sankalan Vyāvakalanābhyām	By Addition and Subtraction pp 14-17, 86-93	
विलोकनम् Vilokanam	By inspection pp 76-77	

The Sutras of Vedic Mathematics

शून्यां साम्यसमुच्चये *Sūnyām Sāmyasamuccaye*	When it is the same, it is nought pp 78-81, 106-7	
परावर्त्य योजयेत् *Parāvartya Yojayet*	Transpose and Adjust p109	
लोपनस्थापनाभ्याम् *Lopanasthāpanābhyām*	By Elimination and Retention pp 108-9	

The other main sutras of Vedic maths are listed below.

आनुरूप्ये शून्यमन्यत्	*Ānurūpyena Śūnyamanyat*	When one is in ratio the other is nought
पूरनापूरनाभ्याम्	*Pūranāpūranābhyām*	By completion and non-completion
चलन कलनाभ्याम्	*Calana Kalanābhyām*	Differential calculus
व्यष्टि समष्टिः	*Vyashti Samashti*	Particular and general
शेषाण्यङ्केन चरमेण	*Śeshañyankena Caramena*	The remainders by the last digit
सोपान्त्यद्वयमन्त्यम्	*Sopāntyamdvayamantyam*	The ultimate and twice the penultimate
एकन्यूनेन पूर्वेण	*Ekanyūnena Pūrvena*	By one less than the one before
गुनितसमुच्चयः	*Gunitasamuccayah*	The product of the sum
गुणकसमुच्चयः	*Gunakasamuccayah*	All the multipliers

These sutras, together with the other sub-sutras will be dealt with in the following books of this series.

Answers

Chapter 1 Nikhilam Multiplication

1.1

1.	81	3.	63	5.	45	7.	9
2.	72	4.	54	6.	36	8.	27

1.2

1.	18	5.	48	9.	54	13.	27
2.	72	6.	63	10.	48	14.	18
3.	64	7.	56	11.	45	15.	9
4.	56	8.	49	12.	36	16.	54

1.3

1.	9215	4.	9408	7.	8463	10.	9024
2.	9409	5.	8930	8.	8645	11.	9215
3.	8439	6.	8648	9.	8924	12.	9212

1.4

1.	8526	5.	8277	9.	7546	13.	9216
2.	8596	6.	6664	10.	8372	14.	8645
3.	7663	7.	6633	11.	6499	15.	8160
4.	8272	8.	8342	12.	7128	16.	8460

1.5

8924	4554	8835	9207
8742	8556	5148	8832
8554			8463
8820			8448
3861	8649	7029	8928

1.6

1.	7426	5.	7268	9.	1862	13.	5244
2.	6555	6.	7189	10.	4753	14.	4277
3.	5664	7.	7410	11.	6460	15.	7826
4.	7347	8.	6900	12.	7238	16.	7125

1.7

1.	6231	5.	6392	9.	6365	13.	3515
2.	6348	6.	7332	10.	5510	14.	6138
3.	7820	7.	6298	11.	7176	15.	3492
4.	7920	8.	7315	12.	6392	16.	4171

Answers

1.8

1.	2574	4.	9212	7.	1188	10.	7546
2.	7227	5.	3861	8.	1683	11.	6664
3.	4752	6.	7350	9.	7663	12.	6499

1.9

1.	9506	7.	9120	13.	9009	19.	8544
2.	9016	8.	8372	14.	8924	20.	8464
3.	8742	9.	8836	15.	8448	21.	8645
4.	9604	10.	9114	16.	7644	22.	6138
5.	5432	11.	7636	17.	4214	23.	6440
6.	7744	12.	7921	18.	3626	24.	4606

Chapter 2 Complements

2.1

1.	36	54	48	01	63	88
2.	76	17	33	52	49	38
3.	67	53	28	44	15	79
4.	55	02	77	66	41	24
5.	07	28	14	35	51	68
6.	56	64	78	62	87	32

2.2

1.	022	235	544	698	234	716
2.	548	692	239	141	226	377
3.	013	131	877	334	295	353
4.	2316	5344	0968	4353	8997	6223
5.	1391	7955	4993	1145	6667	1997
6.	8788	8842	1697	4258	5921	7897

2.3

1.	60	30	840	20	40	10
2.	580	250	610	370	790	220
3.	400	700	600	100	800	300
4.	5350	2320	7850	4530	0920	6950

The Curious Hats of Magical Maths

2.4
1.	88	40	83	84	31	58
2.	496	396	766	499	79	438
3.	0001	877	2346	8902	17	720
4.	4313	6014	41300	88898	39396	69500
5.	6743	80194	3980	24157	06660	57700

2.5
1.	33	5.	232	9.	328	13.	77
2.	7p	6.	18	10.	3235km	14.	Rs2.15
3.	14	7.	Rs24	11.	46	15.	652
4.	143	8.	254	12.	140g	16.	Rs5.80
						17.	990

Chapter 3 - The Deficiency

3.1
1.	21	4.	41	7.	87	10.	46
2.	27	5.	55	8.	78	11.	49
3.	34	6.	37	9.	70	12.	100

3.2
1.	36	5.	37	9.	30	13.	222
2.	47	6.	42	10.	82	14.	385
3.	76	7.	65	11.	252	15.	562
4.	93	8.	97	12.	465	16.	7467

3.3
1.	93	5.	70	9.	55	13.	92
2.	66	6.	64	10.	74	14.	75
3.	87	7.	82	11.	88	15.	96
4.	52	8.	47	12.	34	16.	103

3.4
1.	346	5.	920	9.	637	13.	771
2.	861	6.	215	10.	865	14.	645
3.	276	7.	638	11.	981	15.	822
4.	642	8.	347	12.	802		

Answers

3.5
1. 485
2. 318
3. 747
4. 586
5. 644
6. 387
7. 746
8. 637
9. 415
10. 800
11. 921
12. 943

3.6
1. 18
2. 39
3. 48
4. 65
5. 32
6. 47
7. 175
8. 52
9. 64
10. 333
11. 261
12. 351
13. 528
14. 401
15. 815

3.7
1. 26
2. 4
3. 5
4. 47
5. 336
6. 289
7. 175
8. 724
9. 35
10. 29
11. 23
12. 53
13. 101
14. 426
15. 331

3.8
1. 87
2. 38
3. 117
4. 76
5. 147
6. 158
7. 398
8. 693
9. 236
10. 897
11. 112
12. 196

3.9
1. $6.01
2. 54
3. 491
4. 102
5. 145
6. 702 km

Chapter 4 Number Patterns

4.1
1. 9, 11
2. 10, 12
3. 13, 16
4. 25, 30
5. 20, 24
6. 42, 45
7. 12, 10
8. 27, 32
9. 46, 44
10. 26, 32
11. 44, 53
12. 32, 64

4.2
1. 14, 17
2. 19, 23
3. 11, 16
4. 16, 32
5. 28, 35
6. 75, 70
7. 54, 63
8. 8, 4
9. 42, 47
10. 12, 17
11. 15, 21
12. 25, 36

4.3
1. 17, 21
2. 17, 15
3. 24, 19
4. 16, 32
5. 23, 47
6. 5, 14
7. 45, 56
8. 22, 46
9. 9, 17
10. 15, 42
11. 16, 8
12. 36, 76
13. 12, 11

The Curious Hats of Magical Maths

4.4

1.

2.

3.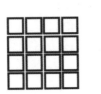

4.

5.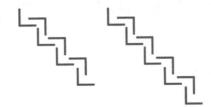

6.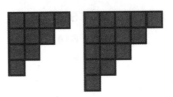

7.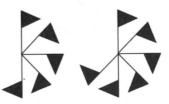

8.

9.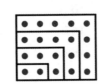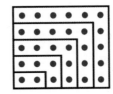

10.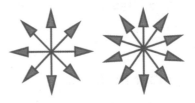

4.5

1. 10   2. 5   3. 60   4. 196   5. 2004
6. Divide the total by 2 and then subtract 2.

4.6

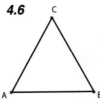 Three guests   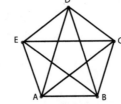 Four guests   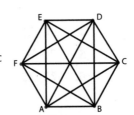 Five guests   Six guests   Seven guests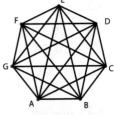

Number of guests	2	3	4	5	6	7	8	9	10	11	12	13
Number of handshakes	1	3	6	10	15	21	28	36	45	55	66	78

4.7

1.
   $1 \times 8 + 1 = 9$
   $12 \times 8 + 2 = 98$
   $123 \times 8 + 3 = 987$
   $1234 \times 8 + 4 = 9876$
   $12345 \times 8 + 5 = 98765$
   $123456 \times 8 + 6 = 987654$
   $1234567 \times 8 + 7 = 9876543$
   $12345678 \times 8 + 8 = 98765432$
   $123456789 \times 8 + 9 = 987654321$

2.
   $1 \times 9 + 2 = 11$
   $12 \times 9 + 3 = 111$
   $123 \times 9 + 4 = 1111$
   $1234 \times 9 + 5 = 11111$

3.
   $9 \times 9 + 7 = 88$
   $98 \times 9 + 6 = 888$
   $987 \times 9 + 5 = 8888$
   $9876 \times 9 + 4 = 88888$

4. 839/2, 629/3, 503/4, 419/5, 359/6, 314/7, 279/8, 251/9. The remainders increase by 1

Chapter 5 Multiplication Above the Base

5.1

1.	132	3.	121	5.	165	7.	144
2.	143	4.	154	6.	187	8.	169

5.2

1.	10812	5.	10908	9.	10403	13.	11021
2.	11118	6.	10914	10.	10504	14.	10609
3.	10404	7.	10706	11.	10710	15.	11449
4.	10807	8.	11016	12.	11130	16.	11336

5.3

1.	11124	5.	10506	9.	10816	13.	10918
2.	10712	6.	10920	10.	11550	14.	11227
3.	10201	7.	11445	11.	11232	15.	11536
4.	10302	8.	11128	12.	11655	16.	10608

5.4

1.	11340	6.	11440	11.	11554	16.	11872
2.	11024	7.	11236	12.	11330	17.	11663
3.	11235	8.	10605	13.	11448	18.	11227
4.	10815	9.	11342	14.	11660	19.	11772
5.	11881	10.	11556	15.	11330	20.	12096

5.5

1.	11413	6.	11433	11.	11322	16.	12566
2.	11544	7.	12096	12.	11424	17.	13493
3.	11760	8.	13464	13.	11984	18.	13390
4.	11628	9.	11730	14.	12240	19.	14382
5.	12342	10.	11449	15.	12360	20.	12750

5.6

1.	13860	6.	13268	11.	14155	16.	15504
2.	14523	7.	15900	12.	15606	17.	13161
3.	13520	8.	16524	13.	14170	18.	13284
4.	13038	9.	17340	14.	14768	19.	13208
5.	15402	10.	12947	15.	13144	20.	16626

Chapter 7 Multiplication by *Vertically and Crosswise*

7.1

1.	156	4.	672	7.	483	10.	961
2.	651	5.	840	8.	484	11.	154
3.	682	6.	660	9.	132	12.	143

7.2

1.	312	4.	325	7.	704	10.	1196
2.	192	5.	546	8.	516	11.	306
3.	714	6.	1271	9.	1722	12.	418

7.3

1.	704	6.	493	11.	770	16.	735
2.	420	7.	378	12.	364	17.	848
3.	416	8.	512	13.	2091	18.	576
4.	546	9.	836	14.	675	19.	1003
5.	1056	10.	322	15.	247	20.	624

7.4

1.	1472	4.	1176	7.	288	10.	336
2.	1104	5.	£848	8.	608	11.	£30.72
3.	4452	6.	378	9.	1113		

Answers

7.5
1. 4402
2. 612
3. 384
4. 744
5. 960
6. 350
7. 2016
8. 336

7.6
1. 67731
2. 28644
3. 40468
4. 25742
5. 59052
6. 67123
7. 80640
8. 81940
9. 85842
10. 52644
11. 55640
12. 93068

7.7
1. 450181
2. 220765
3. 178794
4. 325006
5. 268770
6. 258352
7. 650592
8. 152955
9. 497688

7.8
1. 7650
2. 15808
3. 21462
4. 32860
5. 15582
6. 14868
7. 22072
8. 5640
9. 28050

Chapter 8 Subtraction by *On the Flag*

8.1
1. 24
2. 25
3. 14
4. 29
5. 38
6. 38
7. 43
8. 54
9. 47
10. 28
11. 29
12. 18

8.2
1. 547
2. 339
3. 235
4. 468
5. 456
6. 474
7. 159
8. 158
9. 46
10. 369
11. 152
12. 387

8.3
1. 595
2. 294
3. 299
4. 497
5. 597
6. 296
7. 295
8. 297
9. 99
10. 594
11. 297
12. 98
13. 594
14. 198
15. 393
16. 505

8.4

1.	2025	6.	5450	11.	2965	16.	3997
2.	5297	7.	7299	12.	2693	17.	5297
3.	5237	8.	797	13.	965	18.	1994
4.	3424	9.	5479	14.	2547	19.	6995
5.	4665	10.	6025	15.	1848	20.	997

8.5

1. 274 km
2. 38 cm
3. 352
4. 57

Chapter 9 Nikhilam subtraction

9.1

1.	4	4.	3	7.	9	10.	39
2.	7	5.	6	8.	28	11.	34
3.	38	6.	9	9.	48	12.	9

9.2

1.	69	5.	16	9.	43	13.	63
2.	2	6.	67	10.	8	14.	7
3.	8	7.	8	11.	27	15.	24
4.	28	8.	9	12.	2	16.	8

9.3

1.	289	4.	376	7.	219	10.	489
2.	89	5.	289	8.	86	11.	388
3.	389	6.	388	9.	89	12.	189

9.4

1.	69	5.	87	9.	189	13.	79
2.	89	6.	489	10.	186	14.	88
3.	252	7.	189	11.	386	15.	78
4.	329	8.	268	12.	486	16.	121

9.5

1.	255	7.	150	13.	74	19.	92
2.	460	8.	92	14.	293	20.	91
3.	70	9.	285	15.	85	21.	176
4.	581	10.	122	16.	84	22.	91
5.	362	11.	82	17.	82	23.	92
6.	450	12.	94	18.	21	24.	170

Answers

9.6
1. Rs126
2. 151
3. 226
4. 604
5. 49
6. 58 min
7. 137
8. 362 km

9.7
1. 38445
2. 305825
3. 297264
4. 256145
5. 798051
6. 1399469
7. 3352468
8. 5999980
9. 58819707
10. 38888888
11. 31923407
12. 19889896

9.8
1. 4649
2. 2548
3. 2724
4. 357
5. 3369
6. 4241
7. 593
8. 5140
9. 3947
10. 1025
11. 4187
12. 4635
13. 2364
14. 2955
15. 4327
16. 1607

4378	2724	593	4649	1607	2646
534	9453	3215	7004	2548	2702
3245	8459	1722	3035	4187	1956
4231	4241	4635	357	4327	1133
7558	3055	6811	1719	5140	5634
1947	4558	2310	768	3369	2581
5227	1025	3947	2955	2364	622

Chapter 10 Nikhilam Division

10.1
1. 3/5
2. 2/6
3. 4/6
4. 6/7
5. 5/8
6. 3/6
7. 1/6
8. 8/8
9. 7/7
10. 5/5
11. 3/3
12. 2/7
13. 4/7
14. 2/4
15. 5/7
16. 7/8
17. 4/4
18. 2/8
19. 3/4
20. 4/8
21. 5/6
22. 6/8
23. 1/8
24. 3/8

10.2
1. 11/4
2. 22/6
3. 33/3
4. 12/4
5. 11/5
6. 25/7
7. 15/7
8. 13/6
9. 11/8
10. 15/8
11. 22/3
12. 23/4
13. 35/6
14. 45/7
15. 55/8
16. 68/8
17. 67/8
18. 25/7
19. 58/8
20. 46/8
21. 27/8
22. 46/7
23. 17/8
24. 57/8

125

10.3

1.	124/5	4.	346/7	7.	358/8	10.	222/4	13.	146/8	15.	344/8
2.	137/8	5.	356/7	8.	222/2	11.	356/6	14.	466/6	16.	333/3
3.	234/5	6.	257/7	9.	455/7	12.	233/8				

10.4

1.	5/1	4.	25/0	7.	18/0	10.	46/3	13.	48/1	15.	79/7
2.	6/0	5.	15/1	8.	20/4	11.	35/1	14.	69/3	16.	91/0
3.	9/1	6.	26/3	9.	25/1	12.	29/1				

10.5

1.	1/2	4.	1/3	7.	1/5	10.	1/4	13.	3/7	15.	3/6
2.	12/4	5.	1/7	8.	1/5	11.	1/4	14.	12/6	16.	12/5
3.	2/4	6.	1/5	9.	2/6	12.	2/7				

10.6

1.	2/02	2.	20/40	3.	22/64	4.	224/64	5.	32/56	6.	30/60

10.7

1.	202/02	4.	113/56	7.	114/82	10.	11/54	13.	22/86	16.	521/84
2.	2/22	5.	103/09	8.	12/52	11.	210/50	14.	505/05	17.	212/72
3.	204/8	6.	909/09	9.	113/69	12.	33/63	15.	12/77	18.	98/82

10.8

1.	1/44	6.	2/63	11.	5/82	16.	3/79	21.	8/70	26.	2/49
2.	2/45	7.	5/72	12.	2/73	17.	4/60	22.	4/79	27.	7/88
3.	4/27	8.	1/69	13.	4/69	18.	2/69	23.	3/59		
4.	6/16	9.	3/69	14.	1/55	19.	4/93	24.	2/63		
5.	1/46	10.	8/89	15.	2/78	20.	8/87	25.	2/66		

10.9

1.	3/75	6.	7/66	11.	11/25	16.	115/82	21.	23/76	26.	10/55
2.	2/67	7.	6/79	12.	20/77	17.	213/77	22.	2/60	27.	23/88
3.	1/64	8.	2/78	13.	123/67	18.	12/76	23.	3/47		
4.	2/87	9.	1/56	14.	123/64	19.	213/81	24.	23/95		
5.	9/89	10.	2/05	15.	2/37	20.	1/52	25.	11/71		

10.10

1.	11/424	4.	11/556	7.	44/838	10.	21/452	13.	10/865	16.	22/434
2.	11/435	5.	12/729	8.	3/784	11.	23/802	14.	12/747	17.	2/598
3.	12/615	6.	22/445	9.	1/614	12.	20/828	15.	11/509	18.	21/307

Answers

Chapter 11 Proportionately

11.1
1. 8
2. 10
3. 15
4. 12
5. 9
6. 20
7. 14
8. 18
9. 25
10. 16
11. 24
12. 32
13. 41
14. 28
15. 36
16. 46

11.2
1. 30
2. 60
3. 120
4. 80
5. 210
6. 410
7. 75
8. 250
9. 55
10. 115
11. 285
12. 415
13. 56
14. 188
15. 419
16. 457

11.3
1. 28
2. 42
3. 68
4. 46
5. 80
6. 30
7. 94
8. 106
9. 22
10. 144
11. 70
12. 90
13. 112
14. 148
15. 178
16. 130

11.4
1. 45
2. 70
3. 80
4. 110
5. 90
6. 160
7. 150
8. 120
9. 250
10. 420
11. 220
12. 180
13. 240
14. 190
15. 260

11.5
1. 4
2. 16
3. 28
4. 36
5. 60
6. 14
7. 20
8. 30
9. 38
10. 82
11. 126
12. 50
13. 64
14. 124
15. 86

11.6
1. 248m
2. 25yrs
3. 500
4. Rs12.44
5. 21cm
6. 16kg
7. 484
8. 120m
9. £62
10. 76 km
11. 148 kg
12. 32 cm
13. $540
14. 171 m
15. Rs125
16. £924

11.7
1. 66 kg
2. 16
3. $18
4. 90 km
5. 16
6. £4.50
7. Rs240
8. 450 g
9. 42 min
10. 15
11. 25
12. 4.5 km
13. 8
14. 16 days
15. 3 days
16. 20 min

The Curious Hats of Magical Maths

Chapter 13 Algebra and Equations

Codes: *There is nothing either good nor bad, but thinking makes it so.*

Make secret code and give it to a friend to see if they can break it.

13.1
1. 16
2. 20
3. 18
4. 36
5. 6
6. 3
7. 11
8. 2
9. 20
10. 54
11. 30
12. 20
13. 38
14. 54
15. 25
16. 30
17. 65
18. 7

13.2
1. 9
2. 2
3. 12
4. 17
5. 2
6. 15
7. 5
8. 15
9. 24
10. 2
11. 10
12. 0

13.3
1. 5
2. 7
3. 6
4. 2
5. 1
6. 9
7. 1
8. 7
9. 5
10. 22
11. 28
12. 4

13.4
1. 7
2. 1
3. 11
4. 2
5. 8
6. 1
7. 1
8. 3
9. 0
10. 0
11. 11
12. 7
13. 1
14. 0
15. 18
16. 5
17. 18
18. 12
19. 2
20. 1

13.5
1. $12p$
2. $7a$
3. $2b$
4. $6a$
5. $13x + 3c$
6. $10q$
7. $5a + 2b$
8. $7x + 2y$
9. $3a$
10. $4d - 4$
11. $3x + 5y$
12. $2n - 3$
13. $3x - 2$
14. $9 - x$
15. $7p^2$

13.6
1. 7
2. 9
3. 6
4. 8
5. 6
6. 8
7. 3
8. 3
9. 6
10. 9
11. 4
12. 7
13. 9
14. 3
15. 2
16. 3
17. 2
18. 6
19. 5
20. 2
21. 3
22. 10
23. 3
24. 0
25. 4
26. 7
27. 6
28. 5

13.7
1. 6
2. 10
3. 12
4. 21
5. 4
6. 13
7. 8
8. 3
9. 2
10. 12
11. 60
12. 5
13. 5
14. 5
15. 32
16. 7
17. 5
18. 0
19. 12
20. 25

Answers

13.8
1.	4	**3.**	4	**5.**	14	**7.**	59	**9.**	20	**11.**	5
2.	12	**4.**	8	**6.**	94	**8.**	4	**10.**	4	**12.**	9

13.9
1.	2	**3.**	6	**5.**	5	**7.**	2	**9.**	2	**11.**	0
2.	3	**4.**	5	**6.**	11	**8.**	4	**10.**	5	**12.**	21

13.10
1.	3	**3.**	3	**5.**	1	**7.**	6	**10.**	3	**13.**	2
2.	3	**4.**	5	**6.**	3	**8.**	4	**11.**	4	**14.**	0
						9.	1	**12.**	2	**15.**	2

13.11
1.	8	**3.**	15	**5.**	21	**7.**	54
2.	6	**4.**	12	**6.**	4	**8.**	12

13.12
1.	5	**4.**	2	**7.**	3	**10.**	4	**14.**	11	**18.**	2
2.	1	**5.**	10	**8.**	7	**11.**	3	**15.**	6	**19.**	7
3.	5	**6.**	6	**9.**	2	**12.**	3	**16.**	8	**20.**	2
						13.	3	**17.**	4	**21.**	2

Chapter 14 A few short cuts for multiplying

14.1
1.	1225	**3.**	625	**5.**	225	**7.**	7225
2.	3025	**4.**	2025	**6.**	9025	**8.**	5625
						9.	11025

14.2
1.	202500	**3.**	122500	**5.**	22500	**7.**	90250000
2.	422500	**4.**	302500	**6.**	562500	**8.**	20250000
						9.	72250000

14.3
1.	1224	**3.**	621	**5.**	221	**7.**	624
2.	4216	**4.**	216	**6.**	224	**8.**	9016
						9.	5624

The Curious Hats of Magical Maths

14.4
1. 4221
2. 3021
3. 1216
4. 2024
5. 1224
6. 2021
7. 3024
8. 7216
9. 11016

14.5
1. 253
2. 462
3. 363
4. 561
5. 165
6. 286
7. 891
8. 693
9. 121
10. 550
11. 737
12. 418

14.6
1. 4631
2. 1463
3. 2574
4. 7711
5. 5643
6. 4400
7. 5720
8. 3663
9. 3003
10. 8635
11. 14531
12. 23144
13. 72314
14. 56803
15. 91234

Chapter 15 Digital Roots

15.1
1. 5
2. 8
3. 3
4. 8
5. 3
6. 5
7. 6
8. 5
9. 6
10. 2
11. 5
12. 4

15.2
1. 7
2. 9
3. 8
4. 6
5. 2
6. 5
7. 5
8. 3
9. 1
10. 6
11. 9
12. 2
13. 3
14. 3
15. 9

15.3
1. 1/5
2. 1/8
3. 1/4
4. 1/1
5. 4/7
6. 3/7
7. 3/5
8. 7/5
9. 13/6
10. 27/2
11. 25/6
12. 38/3

15.4 DRs, 4, 8, 3, 7, 2, 6, 1, 5, 9, 4, 8, 3
15.5 DRs, 2, 4, 6, 8, 1, 3, 5, 7, 9, 2, 4, 6
15.6 DRs, 3, 6, 9, 3, 6, 9, 3, 6, 9, 3, 6
15.7 DRs, 5, 1, 6, 2, 7, 3, 8, 4, 9, 5, 1, 6
15.8 DRs, 6, 3, 9, 6, 3, 9, 6, 3, 9, 6, 3, 9
15.9 DRs, 7, 5, 3, 1, 8, 6, 4, 2, 9, 7, 5, 3
15.10 DRs, 8, 7, 6, 5, 4, 3, 2, 1, 9, 8, 7, 6
15.11 4 and 5, 3 and 6, 1 and 8, 2 and 7
15.12 The digital roots are all 9.

15.13
1. 8
2. 2
3. 4
4. 7
5. 3
6. 9
7. 9
8. 8
9. 7
10. 8
11. 7
12. 9
13. 7
14. 4
15. 8
16. 4
17. 8
18. 5

Answers

15.14

	1	4	9	16	25	36	49	64	81	100	121	144
DRs	1	4	9	7	7	9	4	1	9	1	4	9

Chapter 16 Further steps with Nikhilam Multiplication

16.1

1. 995006	5. 991014	9. 990021	13. 997002	17. 988027
2. 995004	6. 988036	10. 986045	14. 988032	18. 998001
3. 987040	7. 985054	11. 992016	15. 982081	19. 982080
4. 990024	8. 991008	12. 985056	16. 992012	20. 981090

16.2

1. 886113	5. 658341	9. 664999	13. 258223	17. 402165	21. 704746
2. 929448	6. 924544	10. 486026	14. 382526	18. 626116	22. 352230
3. 785636	7. 598800	11. 596604	15. 246012	19. 769356	23. 288090
4. 730536	8. 842616	12. 478626	16. 353152	20. 307384	24. 453720

16.3

1. 1005006	5. 1013042	9. 1011024	13. 1008016	17. 1839502
2. 1010024	6. 1014049	10. 1016064	14. 1014045	18. 1929600
3. 1015056	7. 1015054	11. 1008015	15. 1002001	19. 1452094
4. 1012032	8. 1017066	12. 1008007	16. 1725444	20. 1264277
				21. 1305200

16.4

1. 985056	5. 987036	9. 1013036	13. 966064	17. 1025046
2. 992015	6. 1011024	10. 935186	14. 1025084	18. 1077076
3. 1014045	7. 989024	11. 1017070	15. 1035096	19. 497000
4. 998001	8. 984060	12. 940275	16. 495012	20. 1126369
				21. 1246115

Chapter 17 Working with decimals

17.1

1. 5.6	5. 7.1	9. 123.79	13. 911.1
2. 20.1	6. 7.52	10. 172.73	14. 675.5
3. 112.4	7. 0.63	11. 51.15	15. 1660.2
4. 28.1	8. 1.244	12. 92.18	16. 800.2

Chapter 17 Working with decimals

17.1
1. 5.6
2. 20.1
3. 112.4
4. 28.1
5. 7.1
6. 7.52
7. 0.63
8. 1.244
9. 123.79
10. 172.73
11. 51.15
12. 92.18
13. 911.1
14. 675.5
15. 1660.2
16. 800.2

17.2
1. 2.5
2. 23.8
3. 11.8
4. 17.6
5. 41.7
6. 23.9
7. 9.4
8. 30.5
9. 1.65
10. 0.49
11. 166.4
12. 136.8
13. 690.6
14. 2.67
15. 7.65
16. 2.626

17.3
1. 90.24
2. 86.24
3. 8.74
4. 84.63
5. 0.9118
6. 8.526
7. 1.56
8. 0.182
9. 106.08
10. 11.63
11. 0.11648
12. 110.21
13. 961.074
14. 98.1034
15. 9840.55
16. 9.66033
17. 0.981078
18. 0.13161

17.4
1. 7.36
2. 91
3. 36.9
4. 0.876
5. 17.76
6. 0.1944
7. 0.2156
8. 2.201
9. 0.2075
10. 20.24
11. 2.01
12. 0.201

17.5
1. Rs180
2. Rs5166
3. 496.8 km
4. 35.3 m
5. £80.75
6. 532 cm^2
7. Rs4482
8. 27.3 m^2

17.6
1. 1.6
2. 7.3
3. 69
4. 141.9
5. 12.13
6. 11.4
7. 43.7
8. 1.75
9. 206.8
10. 283.5
11. 1695.5
12. 220.1

Answers

Chapter 18 Puzzles and problems

Proportionately

1. 32
2. Wed
3. 1/12
4. 3 : 5
5. 8
6. 35
7. $\frac{56}{67}$
8. 38%
9. 840kg
10. 450ml
11. 108cm
12. 11.24
13. $270 cm^2$

When it is the same, it is nought

1. 10, All letters except S.

2.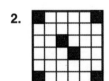

3.

4. Three. Dice 2, 3 and 6.

5. 30 years old

6. 10 and 30 years old

By Elimination and Retention

1. 78
2. 100
3. 8
4. 12
5. $6000 m^2, 12000 m^2$
6. 4

Transpose and Adjust

1. All one half
2. 4 ft and 6 ft
3. 4.5
4. Base X Height
5. 4
6. $\frac{15}{21}, \frac{1}{2}, \frac{1}{4}, \frac{17}{36}$
7. $\frac{341}{512}, \frac{2}{3}$
8. $\frac{1}{10}$
9. 5kg
10. $\frac{2}{9}$
11. 7 cm

133